Oxford Physics Series

General Editors

E. J. BURGE   **D. J. E. INGRAM**   **J. A. D. MATTHEW**

Oxford Physics Series

W. T. WELFORD F.R.S.
PROFESSOR OF PHYSICS, IMPERIAL COLLEGE, LONDON

Optics

SECOND EDITION

Oxford University Press 1981

Oxford University Press, Walton Street, Oxford OX2 6DP

OXFORD LONDON GLASGOW NEW YORK
TORONTO MELBOURNE WELLINGTON
KUALA LUMPUR SINGAPORE JAKARTA HONG KONG
TOKYO DELHI BOMBAY CALCUTTA MADRAS KARACHI
NAIROBI DAR ES SALAAM CAPE TOWN

© Oxford University Press 1976, 1981

First published 1976
Second edition 1981

Published in the United States
by Oxford University Press, New York

British Library Cataloguing in Publication Data

Welford, Walter Thompson
   Optics. – (Oxford physics series).
   – 2nd ed.
   1. Optics
   I. Title
   535   QC355.2   80-41847

   ISBN 0-19-851847-1 Pbk
   ISBN 0-19-851846-3

Typeset by Cotswold Typesetting Ltd., Gloucester
Printed in Great Britain by Billings, Guildford

Preface to the second edition

In revising and enlarging the first edition I was greatly helped by comments and criticisms from several colleagues, but in particular Dr M. E. Barnett, Dr G. A. Brooker, and Dr L. J. Cox; needless to say, they may not recognize the results of some of their criticisms and they are in no way to be held responsible for what is in this second edition. The principal changes are, the addition of a short section on the speed of light, several additional sections on aspects of geometrical optics, some expansion of the chapter on laser light, and, lastly, a whole new chapter on optical light guides. This latter is, of course, pitched at a very elementary level and is not to be regarded as even the briefest introduction to, say, the use of optical fibres for communication; rather, it is intended to illustrate many of the ideas and principles developed in the previous seven chapters in a context which is of wide interest.

Imperial College London　　　　　　　　　　　　　　　　　　　W.T.W.
May 1980

Preface to the first edition

Present-day physics courses are under increasing pressure, on the one hand to keep up with developments in fundamental physics and on the other to cover a broad range of topics appropriate to the interests of students who may never become professional physicists. Thus the time available for optics in the first or second year of an undergraduate course, as for other branches of physics, decreases, and this has influenced my choice of topics in this book; I have been very selective and, as can be seen from the contents list, I have chosen material which is either basic to the development of the optics of the visible spectrum or which has interesting links with other kinds of optics or other branches of physics. Some may be concerned about what is *not* to be found in this book, e.g. measurement of the speed of light, group velocity, standing waves, the envelope function for diffraction gratings, refractometry, Fresnel diffraction, and phase-change effects in interferometry. These omissions might have been dictated anyway by the agreed size of the Oxford Physics Series texts, but I do not plead this as an excuse. The book as it stands is intended as a reasonable selection of topics to be presented to undergraduates, perhaps in their first term at University and certainly having to cope with many other new things at the same time.

I have tried to stress physical arguments, and in order to reduce the mathematical complexity I have introduced the concept of a complex amplitude in the first chapter. I have also used the formalism of Fourier-transform theory freely, since this illuminates and simplifies every branch of physics in which waves appear; this may seem rather extreme for an elementary text, but since simple experiments with lasers are most easily discussed in terms of Fourier transforms it seems almost certain that students will meet the transform in their laboratory work and will grasp the basic ideas even if they have not been presented with a systematic formulation. However, Sections 5.5 and 6.6 contain some more difficult Fourier-transform material, which could be omitted in very early courses. The main definitions and theorems of Fourier-transform theory needed are given, without proofs, in the Appendix.

Some of the problems at the end of each chapter amplify the text by introducing simple extensions of the main discussion.

I should like to thank my colleagues Dr M. E. Barnett and Dr R. W. Smith for their help with this book, mostly given unknowingly; many of their ideas about the teaching of optics have gone into it. Also I am very grateful to Professor E. J. Burge, who read the first draft, gave very valuable criticism, and made many useful suggestions, and to Miss Lesley Harwood, who prepared the index; and I thank the staff of Oxford University Press for their help during publication.

The quotations from James Joyce's *Ulysses* are by kind permission of the Society of Authors, as the literary representative of the Estate of James Joyce, and of The Bodley Head, as publishers.

Imperial College London   W.T.W
1975

Contents

1. Waves, rays, and particles

But what I am anxious to arrive at is it is one thing to invent for instance those rays Röntgen did, or the telescope like Edison, though I believe it was before his time, Galileo was the man I mean. The same applies to the laws, for example, of a far reaching natural phenomenon such as electricity . . .

James Joyce: Ulysses

1.1. The electromagnetic spectrum

For many purposes optics can be regarded as the study of visible light, although in fact this light forms but a small part of a great range or spectrum of radiation. The most familiar part of this spectrum (apart from visible light) is probably the radio region (wireless waves). The complete spectrum of electromagnetic waves is described in Chapter 1 of *Radiation and quantum physics* (OPS 3) by D. J. E. Ingram. The waves are classified according to their wavelength λ or their frequency v and these are related by

$$\lambda v = \text{speed of the wave.} \tag{1.1}$$

Electromagnetic waves of all frequencies have the same speed in vacuum, approximately 3.10^8 m s^{-1}; this universal constant is denoted by c.

We shall begin by describing light and other parts of the electromagnetic spectrum as electromagnetic waves, but this is only one possible description; light (as all other regions of the spectrum) has many properties which are better discussed in terms of other representations (e.g. rays or particles), and we shall have to consider these also.

An electromagnetic wave can be represented as in Fig. 1.1. The graph represents the strength of the electric field in the wave at a given instant and at different points along the direction z of travel. Figure 1.2 shows the same thing in a more picturesque way; the closeness of the lines indicates the relative strength of the electric field. Thus Figs. 1.1 and 1.2 can be regarded as snapshots of the wave in space, taken at a certain instant of time. We could also look at a single point in space and consider the variation in time of the electric field at that point; we should then have a graph like Fig. 1.3.

A more complete picture would be obtained by making the graph of Fig. 1.1 move to the right along the z-axis at the speed c of the wave. The field

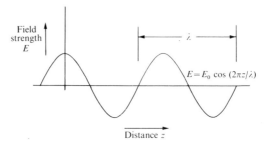

FIG. 1.1.   The electric field strength in an electromagnetic wave at a given instant as a function of the propagation distance z.

FIG. 1.2.   The amplitude of a wave. The closeness of the lines represents the field strength and broken lines indicate negative amplitudes.

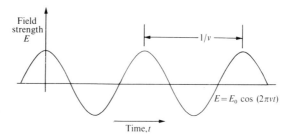

FIG. 1.3.   The electric field strength in an electromagnetic wave as a function of time t.

strength at any point as time passes would then vary as in Fig. 1.3. This travelling wave then has electric field strength E at any distance z and any time t given by

$$E = E_0 \cos 2\pi(vt - z/\lambda). \qquad (1.2)$$

This is easily verified by keeping t or z constant and comparing with the expressions in Figs. 1.1 and 1.3 respectively. A wave travelling to the left, i.e. in the negative z direction, would have a positive sign in the argument of the cosine.

To complete the picture of electromagnetic wave we ought to consider also the accompanying magnetic field. But here it is sufficient to note that the magnetic field has a similar sinusoidal variation and that in the simplest situations, where the wave is not transferring energy to the medium through which it is travelling and where all parts of the wave are travelling in the same direction, the magnetic field varies in step or in phase with the electric field; both fields are at right-angles to the direction of travel of the wave and the electric and magnetic fields are perpendicular to each other.

Different sections of the electromagnetic spectrum are produced and detected in different ways, and the waves have a variety of interactions with matter, (see *Radiation and quantum physics* (OPS 3)). Although we shall be mainly concerned with visible light, it is easiest to consider also the properties of radio waves. This is because many of the properties we shall be interested in—those which produce interference and diffraction effects—can be demonstrated and explained for radio waves with fewer complications than for visible light.

1.2. Power and energy

An essential property of all waves is that they transfer energy (from a source to a detector) without transferring the medium in which the waves occur. Indeed it is doubtful whether there can be said to be a 'medium' for electromagnetic waves. Thus the rate of energy flow or the power in a wave is of interest. It follows from the detailed study of electromagnetic waves that for a wave like that in Figs. 1.1–1.3 the power density (i.e. power per unit area across the wave transmitted in the direction of propagation) is proportional to the square of the electric field strength. We shall take this result as our starting point for a discussion of energy flow; it is treated in detail in texts on electromagnetism (e.g. *Electromagnetism* (OPS 1) by F. N. H. Robinson), where derivations and conditions of applicability can be found. Thus from (1.2) the power density is proportional to

$$E_0^2\{1 + \cos 4\pi(vt - z/\lambda)\}. \tag{1.3}$$

Clearly the cosine term causes a periodic fluctuation in energy flow across a certain plane, say $z = 0$. The oscillating electric field induces an alternating voltage in a conductor (antenna), and this constitutes detection of the electromagnetic wave.

One of the major differences between electromagnetic waves at radio and at optical (and higher) frequencies is that we have no detectors which can respond fast enough to demonstrate optical frequencies directly. In fact the fastest detectors of light will respond only to frequencies of the order of 10^9–10^{10} Hz, some 5 orders of magnitude too low. Thus any detector of electromagnetic radiation in the optical range responds only to the average power over many cycles of the waves. This *time-averaged power* is thus (from eqn (1.3)) proportional simply to E_0^2.

1.3. The complex exponential notation and the complex amplitude

Another basic property of electromagnetic waves is that if two or more wave systems cross in a certain region of space, the electric and magnetic field strengths in this region are found simply by adding as vectors the fields from the individual wave systems. Thus we find the effect of overlapping waves by adding their field strengths or by linear combination. This simple result is not true for very large field strengths, and the topic of *nonlinear optics* has developed in the last decade now that such field strengths are available at optical frequencies. However, in this book, we shall assume linear combination or *superposition*.

Both interference and diffraction phenomena can be explained in terms of superposition of waves, and in this section we shall discuss the mathematical symbolism for this.

Suppose we have two electromagnetic waves of the kind described in section 1.1, travelling at an angle θ to each other, as in Fig. 1.4. Let the two waves have the same frequency (and therefore the same wavelength) and the same maximum field strength E_0. If we use axes as in the figure we can write the two waves as

$$E_0 \cos 2\pi(vt - z/\lambda),$$

$$E_0 \cos 2\pi\{\varepsilon + vt - (z \cos \theta + y \sin \theta)/\lambda\}.$$

(1.4)

In the expression for the second wave the constant ε, known as a phase-shift term, allows for the possibility that the two waves are not in step at the origin of the coordinate system, and the expression $z \cos \theta + y \sin \theta$ ensures that the lines of constant electric field, or wavefronts, are at an angle θ to the y-axis. To fix our ideas we can regard each of the parallel lines in the figure as representing maximum field at a certain instant of time, but this is not essential. In order to find the *interference field*, as it is called, in the region where the waves cross we have to add the two expressions (1.4). If we are dealing with optical frequencies we can only observe the time-averaged power density, which is, of course, what we ordinarily know as the light intensity, and so we have to square the sum of the two expressions in eqn (1.4) and find the time-average. There is no fundamental difficulty in doing this, but the manipulation of the trigonometrical expressions is very involved, particularly if we want to consider more than two waves and if they all have different field strengths. This has led to the introduction of the complex exponential notation and the use of the *complex amplitude* to describe waves, as follows.

First we replace an expression such as that in eqn (1.2) by

$$E = E_0 \exp 2\pi i(vt - z/\lambda),$$

where i is, of course, $\sqrt{-1}$. We shall add complex expressions of this kind in superposing waves, but with the understanding that we are actually concerned only with the real parts. Since real and imaginary always remain

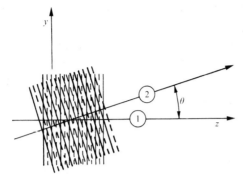

FIG. 1.4.   The superposition of two electromagnetic waves travelling in directions at an angle θ to each other.

separate in summations, this is valid. The above expression represents a wave with plane wavefronts travelling in the z-direction. We can now represent a similar plane wave, travelling in an arbitrary direction specified by a unit length vector $\mathbf{a}$, by the expression

$$E = E_0 \exp 2\pi \mathrm{i}(vt - \mathbf{a} \cdot \mathbf{r}/\lambda),$$

where $\mathbf{r} = (x, y, x)$ is the vector from the origin to an arbitrary point in space. We can check that this agrees with the second of eqn (1.4) by expanding the scalar product and remembering that the components of a unit vector are direction cosines.

Next we put $2\pi v = \omega$, the angular frequency, and we put $2\pi \mathbf{a}/\lambda = \mathbf{k}$. $\mathbf{k}$ is called the wave-vector, and we shall also use the scalar $|\mathbf{k}| = 2\pi/\lambda$, which we denote by k and call the wave-number. Thus our expression for a plane wave is

$$E(t, \mathbf{r}) = E_0 \exp \mathrm{i}(\varepsilon + \omega t - \mathbf{k} \cdot \mathbf{r}). \tag{1.5}$$

We have now indicated explicitly that the field strength E is a function of the position $\mathbf{r}$ and the time t, and we put in an arbitrary phase shift ε. We get the effect of superposing n waves of this kind by adding the appropriate terms,

$$\sum_n E_n \exp \mathrm{i}(\varepsilon_n + \omega t - \mathbf{k}_n \cdot \mathbf{r}),$$

or, taking out the common factor $\exp \mathrm{i}\omega t$, since we have supposed all the waves to have the same frequency,

$$\exp \mathrm{i}\omega t \sum_n E_n \exp \mathrm{i}(\varepsilon_n - \mathbf{k}_n \cdot \mathbf{r}).$$

We can write the summation, which is independent of the time, as $R + \mathrm{i}I$, where R and I are two real functions of the position vector $\mathbf{r}$. From section 1.2

the intensity in the wave-field is the time-average of the square of the real part of

$$(R + iI) \exp i\omega t,$$

i.e., the time-average of

$$(R \cos \omega t - I \sin \omega t)^2.$$

It is easily verified that this time-average is simply $\frac{1}{2}(R^2 + I^2)$. The factor $\frac{1}{2}$ is usually dropped.

In this calculation the time-dependence of the waves appeared as a common factor $\exp i\omega t$ to all terms, which vanished in the final time-averaging; and the final intensity $R^2 + I^2$ is simply the squared modulus of the summed complex expressions.

Thus we have the rule that, to find the intensity due to several superposed plane waves of the same frequency, we add terms of the type $E_n \exp i(\varepsilon_n - \mathbf{k}_n \cdot \mathbf{r})$ for the individual waves and take the squared modulus at the end to find the intensity. An expression of the type

$$E \exp i(\varepsilon - \mathbf{k} \cdot \mathbf{r}),$$

in which the time-dependent part is omitted, is called a *complex amplitude*. These quantities can also be used for superposing other than plane waves (i.e. convergent or divergent waves), and for calculations with all forms of wave motion, not only electromagnetic waves. It is only necessary that the waves all have the same frequency. As a trivial example, the complex amplitude of the wave in eqn (1.2) is

$$E_0 \exp(-2\pi i z/\lambda),$$

and the intensity is therefore immediately E_0^2. If we now apply the procedure to the two waves of eqn (1.4) we easily find, for the intensity in the plane $z = 0$, the expression

$$2E_0^2(1 + \cos\{(2\pi/\lambda)y \sin \theta\}).$$

This is a typical two-beam interference expression; we shall examine it more closely in Chapters 3 and 6.

As we noted earlier, the intensity, which has dimensions of power per unit area, is strictly proportional to E_0^2, i.e. in our present terms it is proportional to the squared modulus of the complex amplitude. The proportionality constant is important both for its dimensionality and for its numerical magnitude in connection with radio wave and microwave theory, but it is not important in the optical problems that we shall encounter. Thus for many purposes we can ignore the electromagnetic nature of light and discuss its properties in terms of a complex amplitude of some undefined quality or medium. Often we need not even specify whether the wave motion is transverse (electromagnetic waves or waves on a string) or longitudinal (sound waves in air). This apparently abstract approach has advantages: parallels with other kinds of wave can be drawn, and we shall find it easier to come to terms with the fact that even the electromagnetic theory is not adequate to explain all optical phenomena.

It is found that all kinds of waves have to be characterized by two different quantities. These are of widely differing physical natures, depending on the kind of wave, but in all cases there is an *amplitude*, which varies in time and space and gives interference effects, and an *intensity*, which represents the rate of energy transport. With suitable interpretations the complex amplitude and its squared modulus, the intensity, can be used in all cases. All interference experiments and many diffraction experiments can be described in these terms.

1.4. Sources and detectors

The production and detection of different parts of the electromagnetic spectrum are described in *Radiation and quantum physics* (OPS 3). Many of the effects and techniques which we usually call 'optical' apply mainly to the infrared, visible, and ultraviolet regions. In these regions there are three main kinds of source:

(1) thermal sources which produce a continuous spectrum, e.g. solid hot bodies, such as filament lamps, and hot gases under high pressure, as in an electric discharge through xenon (e.g. a flash tube);
(2) thermal sources giving line spectra, e.g. mercury vapour or neon discharge tubes under low pressure;
(3) lasers.

The first two are sometimes said to emit *thermal* or *chaotic* light; the latter term reminds us that the phase relationships between light emitted by different atoms or molecules are quite random, whereas in lasers the atoms emit in phase with each other. A continuous spectrum emitted by a source of the first kind may sometimes approximate to *black body* radiation (*Radiation and quantum physics*, OPS 3).

We can describe the production and detection of radio waves quite well in terms of the classical theory of electromagnetism, i.e. without invoking the existence of electrons or using quantum theory. However, in the optical region of the spectrum we have to introduce quantum concepts in order to explain light production and detection, although effects concerned with propagation alone (e.g. interference and diffraction) can be described in terms of a simple wave theory, usually involving only the use of the complex amplitude.

The quantum theory of light emission and absorption is explained in *Radiation and quantum physics* (OPS 3). Here we need only note that electromagnetic radiation is emitted or absorbed in finite quanta of energy called *photons*. The amount of energy in a photon depends on the frequency of the radiation and is given by

$$E = hv = hc/\lambda \tag{1.6}$$

where h, the Planck constant, is 6.626×10^{-34} J s. The energy per photon is sometimes given in electronvolts (eV); 1 eV is 1.602×10^{-19} J. The emission or absorption of a photon corresponds to a change in the energy of an atom, molecule, or other system. In the infrared these transitions are between rotational or vibrational states of molecules; in the visible and near ultraviolet they correspond to changes in the energy levels of electrons in the outer orbits of an atom; and in the far ultraviolet and X-ray regions the inner electrons are involved. These are progressively greater changes in energy of a molecule or atom, and so they produce more energetic photons. This affects the mode of detection. A far ultraviolet photon has enough energy to ionize a gas atom or molecule, and it can be detected by an ionization chamber; alternatively it can cause photoelectric emission of electrons from almost any conductor, and so it can be detected by photodiodes. Visible-light photons can produce photoelectrons from particular surfaces (photocathodes), so that they can also be detected photoelectrically; they can produce a latent image in a photographic emulsion (a complex process which is not yet fully understood), and they can produce photochemical reactions, one of which is the starting point of the process of vision.

We call all the above processes *quantum detection processes*, because they involve a change in state of an individual atom or molecule in a detector by a single quantum. As we move up the wavelength scale into the infrared, fewer quantum detectors are available and the radiation is detected by its general heating effect when it is absorbed, e.g. by a thermocouple or by a resistance thermometer (*thermal detection processes*). The main reason for this difference is that, in the infrared, the amount of energy in a photon is so small as to be comparable to the average random energy of thermal motion of the atoms or molecules in a detector. This thermal energy is of order kT, where k is the Boltzmann constant $(1.381 \times 10^{-23}$ J K$^{-1})$ and T is the absolute temperature. Thus at ordinary temperatures the thermal energy is about $1/40$ eV, and we should not expect to be able to make detectors depending on a quantum effect for photons of energy less than or comparable to this value. We can improve the quantum detection process by cooling the detector, and there are a few devices in which the limitation is circumvented by special procedures, but, broadly speaking, the small photon energy in this region of the spectrum is the reason for the distinction between quantum and thermal detectors.

Detectors vary in their time-response to electromagnetic radiation. Photocells and photomultipliers respond to changes taking place as rapidly as 10^{-9} s; the human eye can only see changes slower than about 0·05 s, an effect known to us as persistence of vision; and photographic detectors can add or integrate light flux for perhaps several hours, until a certain saturation of exposure has been reached. We can also classify detectors according to whether they record images or total flux. Photomultipliers, ionization chambers for X-rays, and thermocouples for the infrared are *total flux detectors*; but the eye, the photographic plate, and television camera tubes record detailed images, i.e. they are *flux density detectors*.

It is often important to know whether a detector is linear in response. The output of a detector can take many different forms, such as a voltage, a current, blackening of a photographic emulsion, or a visual impression in the brain. If we refer to any of these as the 'signal' the detector is said to be linear in response if the signal is directly proportional to the flux falling on the detector. Thus the signal corresponding to the sum of two fluxes is the sum of the individual signals. Photocells and photomultipliers are linear over many orders of magnitude if they are used with suitable circuits, but photographic detectors are usually not. The eye is very non-linear; the sensation of brightness is roughly proportional to the logarithm of the light intensity.

A final interesting property of detectors is their spectral range of sensitivity, or working range. This is roughly indicated in electromagnetic spectrum charts (see *Radiation and quantum physics*, OPS 3), but more information can be provided by a graph. Such a graph may take several forms: (1) we can plot the output signal per unit wavelength interval which the detector would give if used with a fictitious source giving the same energy flux per unit wavelength interval all over the spectrum; (2) we could make a similar plot but per unit frequency interval for detector and source; (3) for a quantum detector we can plot the reciprocal of the average number of photons required to produce one photoelectron as a function of wavelength or frequency, this being called the *quantum efficiency*. Figure 1.5 shows spectral-sensitivity curves for the eye and for some photocells. The difference in response over even the narrow range of the visible spectrum complicates the comparison of responses of different kinds of detectors. For example, it can be seen that if we had beams with equal powers in watts (W), but of violet and green light, an antimony-caesium photocathode would suggest that the violet was the more intense, but an eye would indicate the reverse. This has led to the use of a special system of visual photometric units applicable to the eye in which the *lumen* is the unit of flux (see Welford 1962); a lumen is the equivalent of about 1.467×10^{-3} W of green light or about 1.2 W of violet light of wavelength $0.410\ \mu\text{m}$. In this book we shall not use this visual system of units, since it applies specifically to one detector; light flux will be measured either in watts or in photons per second, with the wavelength or frequency specified.

1.5. Monochromatic and polychromatic fields

We must now consider an essential difference between radio waves and light waves. In sections 1.1 and 1.2 we represented a radio wave as a sinusoidal

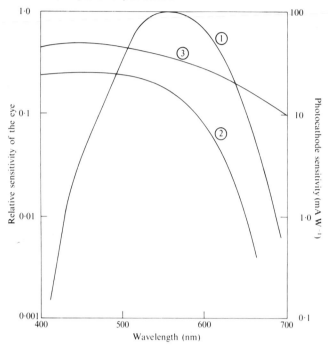

FIG. 1.5.   Spectral sensitivity of the normal eye (curve 1), of an antimony–caesium photocathode (curve 2) and of a sodium–potassium–antimony–caesium photocathode (curve 3). The ordinate scale for the eye, on the left, is in arbitrary units, scaled to unity at the maximum sensitivity. The scale for the photocathodes, on the right, is in milliamperes of photocurrent per watt of incident light power.

variation of electric field (see e.g. eqn (1.2)), so that if a detector with sufficiently rapid response were stationed at a fixed point in the field the output signal would be a strictly sinusoidal or simple harmonic function of time. This is very accurately true for unmodulated radio waves, e.g. the simple carrier wave for radio or television. However, it is not generally the case for visible light. If we could examine the light vibrations carefully over a sufficient number of cycles of the vibrations (and there are indirect ways of doing this), we should find that, although the vibration seems to be simple harmonic for short lengths of time, when examined over longer periods the amplitude varies irregularly and the maxima and minima do not recur at exactly equal time-intervals. Figure 1.6 suggests this effect of random variation of phase and amplitude in a beam of light.

The cause of the randomness in a light beam is easily seen. A radio wave is generated by an alternating current (i.e. a stream of electrons) in a conductor; and it is as if all the electrons are in step all the time, to a very close

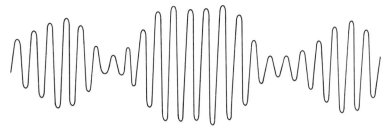

FIG. 1.6.   An almost monochromatic wavetrain.

approximation. However, a beam of light is the summation of a large number of elementary waves (i.e. photons) emitted by atoms or molecules, and in general there is no fixed relationship between the times at which the different photons are emitted. Thus the instantaneous amplitude in the beam of light is obtained as the resultant of many independent waves of random phase but the same frequency.

Under certain conditions we can still regard the light beam as simple harmonic, and then it becomes fairly easy to discuss interference and diffraction. It is for this reason that we said earlier that it is simplest to speak in terms of radio waves first. The two kinds of wave motion we have described may be called *monochromatic* and *polychromatic*.

The actual lengths of time for which light beams can be regarded as simple harmonic vary greatly. At one extreme we have 'white light' from, for example, a tungsten filament lamp. Such light contains a continuous range of wavelengths—as was shown by Isaac Newton, using a prism—and since the wavelengths have random phases we would not expect the simple harmonic property to persist for more than a few periods, i.e. about 10^{-14} s. In a 'monochromatic' beam, e.g. one of the spectrum lines from a mercury discharge lamp, for several reasons the photons have a (small) range of frequencies, and since they are emitted with random phases there is again a finite time involved. Depending on the temperature and pressure of the discharge the time for which the wave persists as substantially harmonic may be some thousands or tens of thousands of periods, i.e. 10^{-11}–10^{-10} s. By means of the speed of light (3×10^8 m s^{-1}) we can express this in terms of the *length* of the wave train which is substantially simple harmonic in form, i.e. a few millimetres to a few tens of millimetres. These times and distances can only be stated approximately, since the wave-trains do not change abruptly but gradually, as in Fig. 1.6. They are determined by experiments with an interferometer (see section 6.2).

Laser light contains even longer stretches of simple harmonic waves. This is because the mode of action of the laser constrains the light-emitting atoms to emit photons which are precisely in phase with each other, instead of having random phase relationships, as in an ordinary light source. There are still some random fluctuations, but in carefully stabilized lasers the light may be truly simple harmonic for times as long as 10^{-6} s, so that the wave-trains are about 1000 m long. The time for which the wave-train remains simple harmonic is called the *coherence time* and the corresponding distance, i.e. this time multiplied by the speed, is the *coherence length*. These quantities determine the possibility of getting certain interference effects. If we have two radio beams from two aerials powered from the same radio-frequency transmitter, there is always the same phase relationship between the electric fields at a certain point. Thus in Fig. 1.7 if A and B are the two transmitters, and if P_1 is 4 wavelengths from A and 5 wavelengths from B, there will be a powerful signal at P_1 in a receiver,

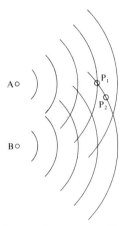

FIG. 1.7.   Interference between coherent waves. The arcs indicate maxima of the amplitude at a given instant. At P_1 the disturbances reinforce at all times, at P_2 they cancel.

since the two waves add in phase. However, at P_2, which is 4.5 wavelengths from A and 5 wavelengths from B, there will be nearly zero signal. It is easy to map regions of maximum and zero signal, and in fact this is the principle of several navigational systems. In optical terms the two sources generate an interference pattern. When a ship or aircraft travels through the interference pattern its changes in position can be found from the maxima and minima traversed; ambiguities can be resolved by, for example, setting up two interference patterns of two different frequencies. However, if the sources are in the optical rather than in the radio-frequency range, the phase relationship between the fields at a certain point is only constant for times comparable to the coherence time. Since the coherence time for ordinary sources is shorter than the response time for the fastest detectors, this means that it is impossible to observe interference between two different light sources. Alternatively, we can say that the interference field varies randomly more rapidly than the response time of the detector and so the detector records a time-averaged interference field. This is why optical interference experiments are always done with light from the same source, split and suitably recombined.

1.6. Waves, particles, and rays

Theories of the nature of light have alternated between those involving waves and those involving particles. For waves the essential feature is the wavefront or surface of constant phase, and for particles it is the ray or particle trajectory. Very persuasive experimental evidence is available for both kinds of theory: interference and diffraction effects for wave theories: quantum phenomena of emission and absorption and the effects of geometrical optics for particle theories.

The present-day view is that we do not understand everything about light (or indeed about any other physical effect), and that in order to obtain theories from which we can make true predictions of experimental results we sometimes have to use wave ideas and sometimes particle ideas. This view has arisen during the present century, from the development of quantum mechanics. According to this theory any particle has wave-like aspects which must be appealed to in order to predict results of some kinds of experiment, and equally, waves have particle-like aspects which must be used in certain cases. In any given optical experiment either the wave or the particle aspect must be stressed.

Figure 1.8 shows the relationships between some of the principal different descriptions or models of light; beside each are listed some of the effects for which the description is used. We can regard these as a series of approximations, starting from the most detailed (and mathematically most elaborate) in terms of photons. Then, if loosely speaking, we allow the Planck constant h to tend to zero we arrive at electromagnetic waves; if we neglect all but one component of the electric vector we obtain scalar waves; if we let the wavelength tend to zero we get rays and geometrical optics; and finally, if we assume all angles are very small, we obtain Gaussian or paraxial optics. We must not, however, regard, say, Gaussian optics as being in any sense 'wrong' because so many approximations were involved. It is simply the right formulation for solving certain problems, e.g. it is easy to get the well-known thin-lens formula

$$\frac{1}{l'} - \frac{1}{l} = \frac{1}{f'},$$

from Gaussian optics but it would be rather tedious to derive it by rigorous quantum mechanics. In this way optics shows clearly that we have no universal physical theory which will explain and predict everything, and that

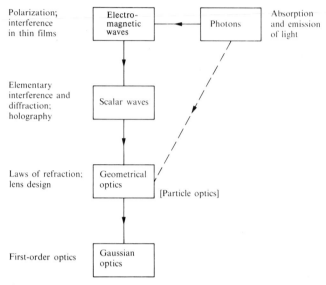

FIG. 1.8.   Descriptions of light.

meanwhile we must be careful to use the appropriate theory for any given problem.

We can compare the wave concept of light with the wave properties of particles such as electrons and protons. In a picture at the same simple level which we have adopted above in comparing the photon and electromagnetic wave pictures, the wave associated with an electron has a complex amplitude which is found as the solution of a certain partial differential equation—the Schrödinger equation. The squared modulus of this complex amplitude represents, to a suitable scale, the probability of finding the electron in any chosen position, by measurements made according to given rules. Thus the wave nature of the electron, or other particle, appears as an uncertainty in locating the particle, in accordance with the Heisenberg uncertainty principle, and the waves associated with material particles such as electrons and protons, so-called *matter waves*, appear as distributions of probability rather than fields of force. In spite of the fact that matter waves thus seem to have less of physical reality than electromagnetic waves they nevertheless produce the standard wave-like effects of interference and diffraction, and thus we can speak, for example, of diffraction of neutrons and interference experiments with electrons.

1.7 The speed of light

The history of the determination of the speed of light starts with an unsuccessful attempt by Galileo; the first measurement came from astronomical observations and thereafter many famous physicists made successively more precise measurements; finally, at the time of writing, an international commission is giving serious consideration to a proposal to *define* the speed of light in vacuum (as 299 792 458 m s^{-1}) in terms of measured lengths and frequencies!

Galileo reported in 1638 that he had stationed two men with shuttered lanterns some distance apart and had attempted to time the interval between directing one man to uncover his lantern and receiving a corresponding signal back from the second man. This we now know could not have succeeded with the simple equipment then available. In 1675 Römer, a Danish astronomer, noticed that the eclipses of the satellites of Jupiter occurred at varying intervals as the planet traversed its orbit round the sun and he deduced that this was due to the finite speed of light; Römer produced the first estimate, some 30 per cent lower than the correct value. Then Fizeau, Foucault, Cornu, Michelson, and others invented successively more accurate methods of timing light signals over terrestrial distances until by about 1940 the value was known to about one part in 30 000. From about 1950 measurements on microwaves and radio waves gave a precision of about one part in 10^5 with good agreement with the optical measurements. More recently developments in lasers and nonlinear optical techniques have made it possible to compare optical and radio frequencies and thus to achieve an accuracy of 2 in 10^{11} in optical frequency determination. Also the wavelength of light can be determined to about one part in 10^9. Thus both wavelength and frequency can be determined much more precisely than speed and so it seems reasonable to *define* c as given by λv.

Problems

1.1. Calculate the wavelengths of electromagnetic waves of frequencies 10^6, 10^9, 10^{12}, and 10^{15} Hz.
1.2. Write down an expression for the complex amplitude of a spherical wave diverging from a point source.
1.3. Calculate the power density in the spherical wave of Problem 1.2, and show that this leads to the inverse-square law.
1.4. Draw graphs of the energy in a photon as a function of (a) wavelength and (b) frequency, using logarithmic scales to accommodate the spectrum from X-rays to radio waves.
1.5. What are the wavelength and frequency of radiation of which the photon energy is of the order of magnitude of the room-temperature thermal energy of atoms?
1.6. Plot a graph of coherence length against coherence time, and mark on it points corresponding to typical sources discussed in Chapter 1.
1.7. Calculate the quantum efficiencies at wavelength 0·45 μm of the two photocathodes of Fig. 1.5.

2. Geometrical optics

He faced about and, standing between the awnings, held out his right arm at arm's length towards the sun. Wanted to try that often. Yes; completely. The tip of his little finger blotted out the sun's disc. Must be the focus where the rays cross.

James Joyce: Ulysses

2.1. The use of geometrical optics

From the point of view of a pure physicist, geometrical optics is a crude approximation for predicting in broad outlines, and with many reservations, how electromagnetic waves behave. It can also predict, with similar reservations, the trajectories of electrons, neutrons, etc.

The applied physicist sees geometrical optics very differently. It is his most important tool for designing many kinds of optical system. Chiefly these are image-forming optical systems for light and for electrons (e.g. optical and electron microscopes and astronomical telescopes), but geometrical optics is essential for some aspects of the design and use of almost any optical system, from a shaving mirror to a single-lens reflex camera. In addition, it is difficult to describe interference and diffraction without using some of the ideas of geometrical optics, such as mirrors and collimators.

The basic concept of geometrical optics is a simple, everyday notion—light travels in a straight line unless it is reflected, according to a law which seems intuitively obvious, or refracted, according to a rather less obvious law. These laws can be verified approximately using very simple apparatus, but the accuracy with which elaborate optical systems work gives us a very precise verification.

2.2. Rays, wavefronts, reflection, and refraction

A ray (of light) is a familiar concept, and we have to carry out rather careful experiments to show that straight-line propagation of light is not exactly true. In geometrical optics we work in terms of rays of light and an associated abstraction, the point source of light, giving (as in Fig. 2.1) a bundle or *pencil* of rays emitted in all directions. We admit that light travels at a known finite

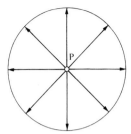

FIG. 2.1.   A geometrical wavefront as a surface of constant phase.

velocity, so that in Fig. 2.1 we can mark on all the rays the points the light reaches in a certain time t after leaving the point source P. If this is supposed to occur in a vacuum or—what is usually assumed to be the same thing in geometrical optics—in air, these points lie on a sphere of radius ct, as shown. If we reverted to a wave picture of light, this sphere would be the surface reached by a wave starting out from P at zero time, i.e. it would be the wavefront. In the present context we are not strictly concerned with waves, but these surfaces can be very useful in geometrical optics, particularly if we consider them after the light has passed through lenses or other optical components. Strictly speaking they are called *geometrical wavefronts*, but we shall usually call them simply wavefronts. Thus a wavefront is a surface reached by the light from a point source in a certain time.

The geometrical wavefronts are almost coincident with the true phase fronts, or surfaces of constant phase according to physical optics, in most regions, but near foci and near the edges of shadows there can be considerable differences, as we shall see in Section 7.1.

Clearly, the rays are normals to the wavefronts in Fig. 2.1. It happens that this is nearly always true in optical systems† although we shall not give the proof of this here, so that we can imagine a system of mutually orthogonal rays and wavefronts propagating through an optical system.

Rays of light go through lenses and mirrors according to the laws of *refraction* and *reflection*. The law of reflection, that the incident and reflected rays make equal angles with the normal to the reflecting surface and are coplanar with it, seems intuitively obvious on almost any theory of light. The origin of this law has not been traced; it was known to Euclid, in about 300 BC. The law of refraction is also concerned with the relationships of the incident and refracted rays with the normal to the refracting surface, e.g. the surface of a sheet of glass. The two rays and the normal are again coplanar, which seems reasonable from symmetry, and the sines of the angles of

† With the exception of certain crystals, but these are not usually used in the kind of optics considered in this chapter.

incidence and refraction have a constant ratio, as in Fig. 2.2.† This ratio depends on the materials on either side of the refracting boundary, and also on the wavelength. The law of refraction was discovered early in the seventeenth century by a Dutchman, W. Snell, and it is therefore called Snell's law.

At first sight the law of refraction seems obscure and *ad hoc*. We may wonder, why sines rather than tangents, or any other function, of the angles? However, in terms of wave theory the form of the law is almost inevitable. Consider a beam of parallel rays striking a plane refracting surface as in Fig. 2.3, and let PP_1 be a wavefront of the incident beam which meets the surface at P at time zero, say. After a certain time t, P_1 has reached the surface at Q_1 and P has travelled on to Q. Thus if the velocities in the two media are v and v', we have

$$P_1Q_1 = vt, \ PQ = v't,$$

or

$$PQ_1 \sin I = vt, \ PQ_1 \sin I' = v't,$$

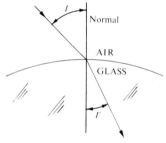

FIG. 2.2.   Snell's law of refraction, $\sin I / \sin I' = $ const.

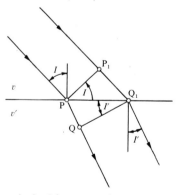

FIG. 2.3.   Snell's law obtained from wave theory.

---

†The law as stated is not true for certain crystals which are anisotropic, i.e. their optical and other properties vary with direction inside the crystal (see Chapter 4).

from which we obtain by eliminating PQ_1,

$$\frac{\sin I'}{\sin I} = \frac{v'}{v}.$$

This is Snell's law. It is usual to put $v/c = 1/n$, where, as in Chapter 1, c is the speed of light in vacuum; n is then called the *refractive index*. Snell's law now takes the form

$$n \sin I = n' \sin I'. \qquad (2.1)$$

The refractive index of a material is a function of wavelength, and for most transparent materials in the visible region it lies between ~ 1.3 and ~ 2.3. The above argument leads also to the law of reflection. At this point we can put them together and for convenience, call them both Snell's law.†

Snell's law can be obtained in an entirely different way, from *Fermat's principle*. This principle states that if light travels from A to B through any optical system it will follow a path such that the time of travel is stationary with respect to neighbouring, but not physically possible, paths. 'Stationary' means that the time of travel may be a maximum or a minimum, or may simply have zero rate of change, as at a point of inflection. The time function at a stationary point could also be behaving like the altitude at the top of a mountain pass, a minimum in some directions and a maximum in others (this is called a saddle point). Pierre de Fermat first stated the principle in 1657 in a form implying that the time of travel is a *minimum* for the physically possible path ('Nature always acts by the shortest course'), but stationarity is strictly more correct. Since the velocity of light in a medium is c/n, where n is the refractive index, this principle can be stated in the form

$$\int_A^B n \, ds \text{ is stationary,} \qquad (2.2)$$

where ds is a differential element of length along any one of the paths from A to B. This is illustrated in Fig. 2.4. The integral of $n \, ds$, which as we saw is proportional to the time of travel of the light, is called the *optical path length*. Thus the optical path length from a point source to all points on a given wavefront is constant.

Fermat's principle is analogous to the *principle of least action* proposed by Maupertuis (1744) as a foundation for mechanics. For a particle in a

---

† This is not generally accepted usage, but it is very easy to put eqn (2.1) into the form of the law of reflection by putting $n' = -n$, as a formal device, for a light ray returning into the first medium after reflection. This then gives reflection as a special case of refraction.

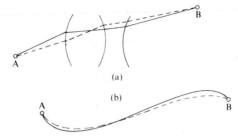

FIG. 2.4.   Fermat's principle. The full line represents a physically possible ray path from A to B and the broken line another path. (a) For lenses. (b) For a medium of continuously varying refractive index.

conservative field of force† the *action* is the integral of momentum along the trajectory, i.e. $\int p \, ds$, and the principle states that for this case the action for a physically possible path is a minimum. We can relate Fermat's principle and the principle of least action by formally making momentum proportional to refractive index. This can be justified for the photon model of light and also, with a suitable definition of refractive index, for electrons.

It is possible to derive Snell's law from Fermat's principle by finding the shortest optical path between points on either side of a refracting boundary (see Problem 2.2). Snell's law can be verified by experiments with prisms. The law is the basis of optical design of all lens systems, and the fact that they work as designed provides extensive experimental verification of the law and thus of Fermat's principle.

If we apply Snell's law (eqn (2.1)) to a ray passing from glass of refractive index n to air we find that at an angle of incidence in the glass I_c, given by $\sin I_c = 1/n$, the refracted ray 'emerges' at an angle of 90° to the normal; for larger angles of incidence eqn (2.1) gives $\sin I' > 1$, and it is found experimentally that the light is completely reflected at the boundary—so-called *total internal reflection*. This follows rather simply from the electromagnetic theory. The angle I_c is called the *critical angle*. Total internal reflection is used in reflecting prisms, of which the simplest and best known is the right-angle prism, as in Fig. 2.5. Light is reflected internally with 100 per cent efficiency whereas at silvered or aluminized reflecting surfaces there is always some loss by absorption of the incident light. The determination of the critical angle is the principle of certain *refractometers*, i.e. instruments for measuring refractive index.

† In a conservative field the work done on a particle in transporting it from A to B is independent of the path from A to B. The force on the particle is the gradient of a scalar single-valued potential. Gravitational and electrostatic fields are examples of conservative fields, but a magnetic field is not conservative.

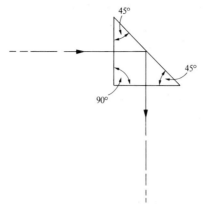

45°

45°

90°

FIG. 2.5.   A right-angle prism; there is total internal reflection at the hypotenuse face.

2.3. Optical images with a thin lens

The formation of images in the optical region by lenses and mirrors is familiar, and it can be demonstrated in many other regions of the electromagnetic spectrum. A simple explanation is as follows. Figure 2.6 shows a plano-convex lens. Light from a point source O on the axis, e.g. a pinhole in metal foil with a lamp behind it, produces a diverging pencil of light with spherical wavefronts convex to the lens. The lens is thicker at the axis.

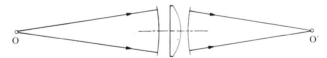

O   O′

FIG. 2.6.   Formation of an image by a lens.

Thus the optical path length through it is greatest there, since the refractive index of the glass is greater than that of air, and the wavefronts transmitted by the lens will be retarded at the centre relative to the edge. Suppose for simplicity that the emergent wavefronts are also spherical in shape, as indeed they must be by symmetry for a small enough diameter lens. Then they may be convex with a longer radius of curvature, because of the greater delay at the centre, or, if the shape of the lens and the position of O are suitable, they may become concave, as indicated. Drawing the rays as normals to one of these wavefronts, we see that they intersect at some point O′ on the axis—this is the image of O.

We can put this in an alternative but entirely equivalent way as follows. If O and O′ are to be object and image then all rays which enter the lens from O must pass through O′. But Fermat's principle (Section 2.2) states that all those ray paths must be such that the optical path length from O to O′ along them is stationary, so that the optical path length from O to O′ must be the same along all the rays. Thus the greater glass path through the centre of the lens compared to the edge is compensated by the shorter length of the rays in air at the centre.

If the object point O is not on the axis a similar argument will show that an image point is again formed, with suitable approximations. Thus we get an image of an *extended object*.

It remains to obtain simple formulae to describe image formation, and this is most easily done by considering that typically optical abstraction, the *thin lens*. This is a lens (as in Fig. 2.7) of refractive index n and with curvatures (i.e. reciprocal of radius of curvature) c_1 and c_2 of the refracting surfaces. We

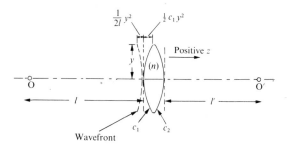

FIG. 2.7.   Formation of a point image.

neglect the thickness of the lens. This seems paradoxical for the lens in the figure, which is biconvex, but the approximation gives useful results. We assume that an object point O at a distance l from the lens produces an image at O′, a distance $l′$ from the lens, and we want to find $l′$ in terms of l.

We have to choose signs for these lengths, and to facilitate this we set up a coordinate system with origin at the lens (it does not matter exactly where in the lens since its thickness is negligible) and z-axis along the optical axis. Then *in our figure* O′ has a positive z-coordinate so that $l′$ is a positive number, and O has a negative coordinate, so that l is negative. We can use the same system to give signs to the curvatures, since the equation of a sphere passing through the origin and with its centre to the right of the origin is of the form

$$z = \tfrac{1}{2}c(x^2 + y^2) + \ldots ,$$

with positive c. Thus in our diagram it happens that c_1 is positive and c_2 negative. We have now settled the problem of signs according to the ordinary conventions of coordinate geometry, and we can use the symbols as in coordinate geometry, i.e. without further concern for signs.

At a distance y from the axis the lens will be $\frac{1}{2}(c_1 - c_2)y^2$ thinner than at the centre, by the above formula, i.e. this amount of glass path is replaced by air, so that the optical path through the lens between two planes tangent to the surfaces will be shorter here than at the centre by $\frac{1}{2}(n-1)(c_1 - c_2)y^2$. The depth of curvature of the incoming wavefront is, as in the figure, $y^2/2l$, and that of the emergent wavefront is $y^2/2l'$, so that to ensure that optical path lengths (i.e. times of travel between corresponding points of two wavefronts) are equal we must have

$$\frac{1}{2l}y^2 + \frac{1}{2}(n-1)(c_1 - c_2)y^2 = \frac{1}{2l'}y^2$$

or

$$\frac{1}{l'} - \frac{1}{l} = (n-1)(c_1 - c_2). \tag{2.3}$$

This *conjugate distance equation* relates the positions of object and image points, which are said to be *conjugates*. As we have derived the result, the effect of the lens is to add to the incoming wavefront an increment of curvature of amount $(n-1)(c_1 - c_2)$. If the incoming rays are parallel, i.e. the incoming wavefronts are plane or the object point is at infinity, the image distance l' is given by

$$l' = \frac{1}{(n-1)(c_1 - c_2)},$$

as in Fig. 2.8. This distance is called the focal length f, its reciprocal K is the *power* of the lens and the image point is called the second principal focus (any

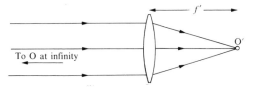

FIG. 2.8.   An object at infinity is imaged at a principal focus.

point where rays meet is a focus). Similarly, if the object point is at a distance l given by

$$l = \frac{-1}{(n-1)(c_1 - c_2)} = -f,$$

the emergent rays are parallel, as in Fig. 2.9. Thus the point at infinity and a principal focus are object and image conjugates. These are two important

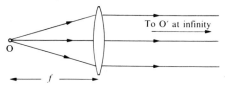

FIG. 2.9.   A collimator. The object at a principal focus produces an image at infinity, i.e. a collimated beam.

cases for physical optics applications. The second is used, as a *collimator*, to produce a beam of plane waves from a point source; and the first is used to show a diffraction or an interference effect which is nominally formed at an infinite distance—in the *far field*—at a convenient place for observation (see Section 3.4).

If the lens forms an image of a small object, say a line segment of height η, as in Fig. 2.10, the image will have a height η', and the ratio of these is the magnification m,

$$m = \eta'/\eta. \tag{2.4}$$

Let α and α' be the semi-angles of the cones of rays forming the image, and let l and l' be the conjugate distances. The ray from the end of the object through

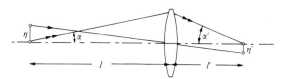

FIG. 2.10.   The image of an extended object. In this figure α and η are positive quantities and α' and η' are negative.

the centre of the lens passes through the lens undeviated, since the lens has its surfaces parallel at the centre, and so we have

$$\eta/l = \eta'/l'$$

or

$$m = l'/l. \tag{2.5}$$

Also, if y is the height at which the other ray shown meets the lens,

$$\alpha = -y/l, \alpha' = -y/l',$$

so that, from eqn (2.5),

$$m = \alpha/\alpha'. \tag{2.6}$$

(2.5) and (2.6) are correct for all signs of the variables, as explained in the caption to Fig. 2.10. In all the above we have been tacitly using the Gaussian or paraxial approximation of small angles mentioned in Section 1.6.

If the object is at the first principal focus, i.e. the focus on the object side, as in a collimator, the rays from each point of the object form a parallel pencil, α' is zero, and eqns (2.4) and (2.6) are not applicable. In this case we say that the image is formed at infinity and we use its angular subtense as a measure of its size. Thus if the object height is η (as above) and if the focal length is f, the ray from the end of the object through the centre of the lens emerges at an angle η/f to the axis, since it is undeviated, and all rays from this point therefore make this angle with the axis in the image space. Thus we have the rule that a point on the focal plane a distance η from the axis produces plane parallel wavefronts travelling at an angle η/f to the axis (collimator). Conversely plane parallel wavefronts travelling at an angle β to the axis form a point image at the focal plane a distance βf from the axis (objective).

2.4. Multi-element lenses

The above ideas can be generalized to lens systems which contain a combination of several more-or-less thin lenses, (e.g. the type of system used in a camera). From eqns (2.5) and (2.6) the magnification depends on the positions of object and image, so it is reasonable to suppose that for a multi-element or thick system we can find a pair of conjugate planes with magnification unity. In most practical cases these will be virtual conjugates, i.e. inside the system as in Fig. 2.11, where they are denoted by P and P', but

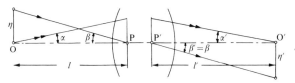

FIG. 2.11.   The principal points of a thick lens, P and P'.

this is not important. These are known as the *principal planes*, and the axial points are the *principal points*. Obviously, for a thin lens the principal points coincide at the lens.

Another ray from an axial object point O to its image O' passes through the two principal planes at the same distance from the axis, since they are planes of unit magnification. Now we can compare Figs. 2.10 and 2.11 for a thin and a thick lens. The figures are similarly labelled to indicate their essential similarity, but in the thick lens there is a kind of limbo or missing space between the two principal planes. However, we can measure l from the first principal plane to the object and l' from the second principal plane to the image, and all will correspond in the two cases. For example, the focal length is the distance from the principal plane to the point where rays from infinity focus (as in Fig. 2.12), and the magnification is again given by $\alpha/\alpha' = l'/l$. Corresponding to eqn (2.3), there will be a conjugate distance equation,

$$\frac{1}{l'} - \frac{1}{l} = \frac{1}{f},\qquad(2.7)$$

but now f, the focal length, depends on the detailed construction of the lens system.

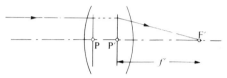

FIG. 2.12.   The principal focus F' and the principal planes for a thick lens.

2.5. Paraxial raytracing

The positions of the principal planes of a multi-element optical system can be found as in Fig. 2.12; since the planes through P and P' are planes of unit magnification a ray entering the system and meeting the plane through P at a certain distance from the axis must emerge from the plane through P' at the same distance from the axis. Then if this ray entered parallel to the axis the point at which the entering and emerging segments meet must be on the plane through P'; similarly the plane through P could be found from the path of a ray entering parallel to the axis from the right. Thus to find P and P' we have to trace the path of a ray through an optical system in the paraxial approximation.

Paraxial raytracing is a step-by-step or iterative process in which the path of a ray is calculated through each surface in turn. We consider a ray, originally from a point on the axis in the object space, and we suppose we have found that the ray meets a certain surface at a distance y from the axis and that its intersection length is l, as in Fig. 2.13. The refracting surface has curvature c and the refractive indices on either side are n and n' as shown. Let the refracted ray meet the axis at a distance l'. Then to find l' we proceed as in Section 2.3, i.e. we find l' by the condition that the optical path length from the (intermediate) object point to its image through the refracting surface shall be the same for all rays, i.e. for the path along the axis and for the ray shown. We draw in the spherical wavefronts touching the refracting surface and then we can write down this condition, after subtracting the radii of the wavefronts, as follows,

$$\tfrac{1}{2}ny^2c - \tfrac{1}{2}ny^2/l = \tfrac{1}{2}n'y^2c - \tfrac{1}{2}n'y^2/l'$$

or

$$\frac{n'}{l'} - \frac{n}{l} = (n' - n)c.\qquad(2.8)$$

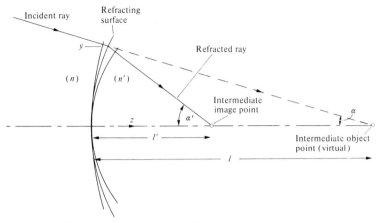

FIG. 2.13.   Notation for paraxial raytracing.

This is a conjugate distance equation for a single surface and the similarity with eqn (2.3), the conjugate distance equation for a thin lens, should be noted.

For numerical raytracing it is more convenient to put this result in terms of the convergence angles α and α'; multiplying eqn (2.8) through by y we have

$$n'\alpha' - n\alpha = -(n' - n)cy \qquad (2.9)$$

Now let the next refracting surface be a distance d along the axis as in Fig. 2.14; then from the figure the new incidence height, y_{+1} say, is given by

$$y_{+1} = y + \alpha'd \qquad (2.10)$$

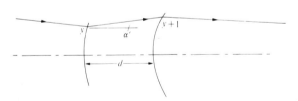

FIG. 2.14.   Transferring from one surface to the next in paraxial raytracing.

We are now ready to use eqn (2.9) again, since the α' from eqn (2.9) is the incoming convergence angle for this next surface; thus the process is taken through the system to obtain α'_n and y_n for the final, n^{th} surface. Paraxial raytracing is usually done numerically since the algebraic eliminations needed to express α'_n and y_n in terms of α_1 and y_1 are cumbersome; however, some simple cases are given as examples at the end of the chapter.

Now returning to Fig. 2.12 we can see that if the incoming ray was parallel to the axis, i.e. coming from the axial point at infinity, the focal length must be given by $-y_1/\alpha'_n$; also the distance from surface n to the focus, sometimes called the *back focal length*, is $-y_n/\alpha'_n$ and the position of P′ is obtained from the difference of these two quantities. Similarly by reversing the process the focal length on the object side and the position of P can be found. The two focal lengths are numerically equal when, as is usually the case, the object and image spaces have the same refractive indices, but not otherwise, as will be seen in Section 2.7.

The paraxial or Gaussian properties of an optical system, i.e. the positions of pairs of conjugate points and the corresponding magnification, are known once the principal planes and foci are found, since we can then use the conjugate distance eqn (2.7), measuring l and l' from the appropriate principal planes. Thus the optical system can be replaced by a skeleton consisting of the principal planes and foci, as in Fig. 2.15, and this can be used

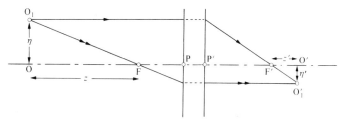

FIG. 2.15.   Graphical construction for finding the image O′ O′$_1$ of an object O O$_1$.

in a simple graphical construction for conjugates. Given an object OO$_1$ we can take a ray parallel to the axis from O$_1$ (single arrow) which must travel across the space between the principal planes at constant height and then in the image space pass through the focus F′. A second ray from O$_1$ (double arrow) through F emerges parallel to the axis in image space and the intersection of these two rays must be O′$_1$, the image of O$_1$.

Figure 2.15 is, of course, a graphical representation of eqn (2.7); it can be used to derive another useful form of conjugate distance equation. Recalling that distances are taken with signs according to coordinate geometry, the image space focal length $P'F' = f'$ is positive in the figure; similarly the object space focal length $PF = f$ is negative as drawn. For complete generality we assume these to be numerically unequal. Let $F'O' = z'$, $FO = z$, again with appropriate signs, so that in the figure z' is positive and z is negative. Then by similar triangles in the figure it can be shown that

$$z/f = -\eta/\eta' = -1/m,$$

$$z'/f' = -\eta'/\eta = -m,$$

so that

$$zz' = ff'. \qquad (2.11)$$

This is Newton's conjugate distance equation, relating conjugates measured from the foci rather than the principal planes.

The raytracing equations as derived above appear to allow only for lens surfaces in the optical system. It is easy to allow for a convex or concave mirror by the formal device mentioned in Section 2.2 of putting $n' = -n$ for a mirror. We then obtain from eqn (2.8) the following conjugate distance equation for a mirror of curvature c,

$$1/l' + 1/l = 2c \qquad (2.12)$$

or, in terms of the convergence angles

$$\alpha' + \alpha = 2c \qquad (2.13)$$

These can easily be verified by a direct calculation as above for eqn (2.8).

2.6. The Lagrange invariant and the power transmitted by an optical system

We recall, from the Section 2.3, the magnification formula

$$m = \eta'/\eta = \alpha/\alpha',$$

where α and α' are the ray convergence angles and η and η' are object and image heights. From this we have

$$\eta'\alpha' = \eta\alpha$$

as a relationship between object and image quantities. We can generalize this to intermediate images formed between lenses in an optical system (as in Fig. 2.16), for the image formed by the system to the left is the object for the next part, and they share the same convergence angle α. Thus $\eta\alpha$ is the same in all air spaces in the system.

Now suppose a plane surface of glass of refractive index n is placed immediately after one of these intermediate images, as in Fig. 2.17. The image height is unaltered if the image is at the surface, but the convergence angle

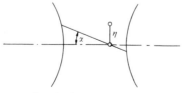

FIG. 2.16.   The Lagrange invariant.

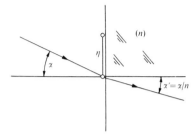

FIG. 2.17.   An intermediate image formed at a plane surface.

becomes $\alpha' = \alpha/n$ in Gaussian approximation. Thus finally we have that the quantity

$$H = n\alpha\eta \tag{2.14}$$

is an invariant throughout an optical system. In this expression η is the intermediate image size corresponding to the original object size, α is the intermediate convergence angle of the ray from the original axial object point, and n is the refractive index. The quantity H is usually called the *Lagrange invariant*; it was discovered independently by several people including Lagrange. It follows that for a system with refractive indices n and n' in the object and image spaces the magnification is given by the statement of the Lagrange invariant:

$$n\alpha\eta = n'\alpha'\eta' \tag{2.15}$$

As well as relating magnification and convergence angle, the Lagrange invariant is a measure of the light flux or power transmission capability of an optical system. In a real optical system the convergence angle α on the object side is usually determined by the size of an iris diaphragm or aperture stop in the system (as in Fig. 2.18), as well as, of course, the distance of the object. Thus a small circular object of radius η mm on the axis will radiate into the lens inside a cone of half-angle α, and if the light power per unit area and per unit solid angle (luminance if it is visible light or radiance for any wavelengths) is B W mm^{-2} sr^{-1}, say, the power collected by the lens will be

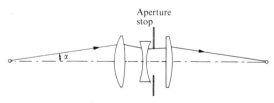

FIG. 2.18.   An aperture stop in a system determines the angle α of the accepted cone of rays.

$\pi^2 B\eta^2\alpha^2 W$. If we ignore attenuation by reflection losses, absorption, and scattering, by conservation of energy this same power flow must occur across any surface of the lens system and across the final image surface. Thus on comparing the above expression with eqn (2.14) we see that the power transmitted by the optical system is proportional to the square of the Lagrange invariant.

The brightness or luminance of an image formed by an optical system is the light power per unit area and per unit solid angle in the image. It is clear from the Lagrange invariant that the luminance of the image is equal to that of the object multiplied by a factor less than unity which allows for attenuation, so that no image can be brighter than the original object if object and image are in media of the same refractive index. If the indices are different this statement has to be modified in an obvious way since the refractive index appears in the Lagrange invariant.

2.7. The relation between the two focal lengths

It was stated in Section 2.5 that the object space and image space focal lengths are not numerically equal if the refractive indices in the two spaces differ. We now use the Lagrange invariant to obtain the relation between the focal lengths. Figure 2.19 is a diagram similar to Fig. 2.15 but with the object and image at the principal planes, so that the magnification is unity. Then the convergence angle α for the object PP_1 can be seen to be $-PP_1/f$ and that for $P'P'_1$ is $P'P'_1/f' = PP_1/f'$. Thus eqn (2.15) takes the form

$$-n\frac{(PP_1)^2}{f} = n'\frac{(PP_1)^2}{f'}$$

so that the required relation between the focal lengths is

$$-\frac{n}{f} = \frac{n'}{f'} \qquad (2.16)$$

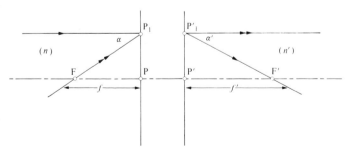

FIG. 2.19. Construction for the relation between the two focal lengths of an optical system.

2.8. Non-paraxial optics

The results of Sections 2.3 to 2.7 are all based on Gaussian optics, according to which the lens and mirror apertures are supposed to be very small and the rays all make small angles with the optical axis. It is possible to define 'small' more precisely, as a mathematical order of magnitude (Welford 1962), but here we shall simply note that experimentally we do find well-defined images under Gaussian conditions but if we go beyond a certain range of angles the images look poorer (e.g. the image of a lamp formed by a convex spectacle lens tipped obliquely, so that the object and image are some way from the lens axis, is not sharp). Under such conditions we cannot rely on the simple approximate equations of this chapter and we have to trace rays exactly, i.e. according to Snell's law. We then find that point objects do not form point images, i.e. there are *aberrations*. Optical systems such as camera lenses and microscopes have many lens components arranged so as to correct or compensate aberrations.

Figure 2.20 shows how one kind of aberration arises when we use a thin

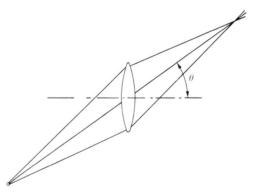

FIG. 2.20.   Oblique refraction by a thin lens. The rays in a section at right-angles to the diagram focus at a greater distance from the lens than those drawn.

convex lens at an angle θ to its axis. We saw in Section 2.3 that a thin lens forms an image of a point object by adding an increment of curvature to the incident wavefront equal to $1/f$, the reciprocal of the focal length; this happens because there is more glass at the centre than at the edge, so the wavefront is retarded more at the centre. In Fig. 2.20 this effect will still happen in the section perpendicular to the plane of the diagram, but in the plane of the diagram the width of the lens presented to the wavefront is less by a factor $\cos \theta$, and since there is, to a sufficient approximation, the same variation in thickness between centre and edge, the increment in curvature will be greater by a factor $1/\cos \theta$. Thus the refracted wavefront will have a greater curvature in the plane of the diagram than in the perpendicular section, and so the rays (normals to the wavefront) will not focus to a single point. This aberration is called *astigmatism.*

Another example of an aberration is *chromatic aberration.* The refractive index depends on the wavelength of light, for all material media—an effect called *dispersion.* For example, Fig. 2.21 shows the refractive index as a

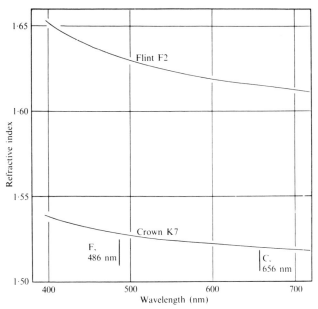

FIG. 2.21.   Dispersion curves of two optical glasses. The names and symbols of the glasses have no physical significance.

function of wavelength for two common optical glasses. The brightest part of the visible spectrum lies between the two Fraunhofer spectrum lines C and F, and it can be seen that over this range the refractive index of the crown glass varies by about 0.008. Since the focal length of a thin lens (see Section 2.3) is given by $1/f = (n - 1)(c_1 - c_2)$, it can be seen that the focal length of a thin lens made of this glass would vary by about $0.008/0.52 = 1.6$ per cent over this wavelength range. This effect is a form of chromatic aberration. Fortunately, as can be seen from the figure, the flint glass has about twice this variation of refractive index over the same wavelength range, and it is thus possible to combine a converging lens of crown glass with a weaker diverging lens of flint glass, so as to cancel the chromatic aberrations but yet leave some converging power. Such a combination, as in Fig. 2.22, is an *achromatic doublet.* We can

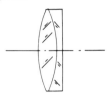

FIG. 2.22.   An achromatic doublet.

choose the powers of the components of the doublet as follows. For a single thin lens the power K is given by

$$K = (n - 1)(c_1 - c_2).$$

Let the chosen wavelength range correspond to a change δn in refractive index. Then we have for the change in K by differentiation

$$\delta K = \delta n(c_1 - c_2)$$

or

$$\delta K = \frac{\delta n}{n - 1} \cdot K \qquad (2.17)$$

Now let the powers of the two components of the doublet be K_1 and K_2, let their refractive indices be n_1 and n_2 and let the required total power of the doublet be K. From Problem 2.8 we have

$$K = K_1 + K_2 \qquad (2.18)$$

and from eqn (2.17) we have for the change with wavelength of the total power

$$\delta K = \frac{\delta n_1}{n_1 - 1} K_1 + \frac{\delta n_2}{n_2 - 1} K_2$$

If we require the power to be the same for the two wavelengths at either end of the chosen range we set $\delta K = 0$ and we then have

$$\frac{\delta n_1}{n_1 - 1} K_1 + \frac{\delta n_2}{n_2 - 1} K_2 = 0 \qquad (2.19)$$

It can be seen that eqns (2.18) and (2.19) are a pair of simultaneous equations for K_1 and K_2, giving the required solution for the achromatic doublet. There is still a choice of curvatures for the individual thin lenses and this is used to correct other aberrations. For a treatment of aberrations in general see W. T. Welford, *Aberrations of the symmetrical optical system* (1974).

2.9. Afocal systems

In our discussion of optical systems we have assumed tacitly from Section 2.3

on that if the object is at infinity the system will form an image at some finite distance. *Afocal systems* form images at infinity of objects at infinity. Passing over the trivial examples of plane mirrors and plane-parallel glass plates, which are certainly afocal, a non-trivial example is the Galilean telescope or opera glass, shown in Fig. 2.23. This system is theoretically adjusted so that the entering and emerging beams are both collimated, so that an object at

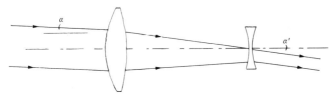

FIG. 2.23.   The Galilean telescope, an example of an afocal system.

infinity is imaged at infinity. Afocal systems fall awkwardly outside the formalism so far developed but this difficulty can be overcome by obtaining an expression for the Lagrange invariant in a space where the object (image) is at infinity. We call the angular subtense of an object at infinity β, i.e. this is the angle which the parallel rays make with the axis as in the figure. Then for l very large but finite we can put $\beta = \eta/l$ and $\alpha = -y/l$, so that from eqn (2.14) we have as l tends to infinity,

$$H = -ny\beta \qquad (2.20)$$

in a space where the object is at infinity. We now define the magnification as the ratio of angular subtenses of the object and the image, $m = \beta'/\beta$, so that for $n = 1$ on both sides of the system

$$m = y/y' \qquad (2.21)$$

i.e. the magnification is the ratio of the diameters of the entering and emerging beams (provided there are no internal apertures to confuse this calculation). We cannot define a focal length or power for an afocal system.

Problems

2.1   Draw accurately rays refracted from air to glass ($n = 1.53$) at angles of incidence from $0°$ to $90°$, at $10°$ intervals.

2.2.   Points P_1 and P_2 are in media of refractive index n_1 and n_2 separated by a plane boundary. Calculate the optical length from P_1 to P_2 along straight-line segments from P_1 to a point Q on the surface and from Q to P_2. By differentiation prove Snell's law.

2.3.   Sketch typical object and image positions, indicating suitable rays, for (a) a camera, (b) a slide projector, and (c) a burning glass.

2.4.   Show that the optical path length along all rays between two wavefronts of a pencil is constant.

2.5. Plot a graph of l' as a function of l for a lens of focal length 100 mm, allowing the conjugates to range between -1000 mm and $+1000$ mm.

2.6. What is the physical significance of positive values of l and negative values of l' in the above example?

2.7. A thin lens has curvatures c_1 and c_2 and refractive index n. Show that the analogue of eqn (2.9) for the thin lens is

$$\alpha' - \alpha = -yK$$

where K is the power of the lens, defined as the reciprocal of the focal length (Section 2.3).

2.8. Two thin lenses of powers K_1 and K_2 are separated by a distance d. Show that the power of the combined system (reciprocal of the focal length) is

$$K = K_1 + K_2 - dK_1K_2.$$

Hence show that for any number of thin lenses in contact the total power is the sum of the individual powers.

2.9. A thin lens has focal length 50 mm. What is the magnification for the following object conjugates: 20 mm, -100 mm, $+100$ mm.

2.10. What is the focal length of a concave mirror of radius r? Calculate and draw to scale the image positions and magnifications for a concave mirror of radius 100 mm for object distances 25 mm, 100 mm, 200 mm.

2.11. A lamp filament in the form of a flat ribbon of area 10 mm^2 radiates 10 W of light. (a) What is its luminance, and (b) how much power is collected by a lens of diameter 30 mm which is 100 mm from the filament?

2.12. Show that a prism with small angle α deviates light rays through an angle $(n-1)\alpha$, where n is the refractive index of the prism.

2.13. A right-angle prism as in Fig. 2.5 is made of material of refractive index 1.52. A beam of light is reflected at the hypotenuse face at the critical angle of incidence. Calculate the angle between the incident and emergent rays outside the prism.

2.14. A Galilean telescope is to be made up of two thin lenses of focal lengths 50 mm and -20 mm. Find the separation between the lenses and calculate the magnification.

3. Propagation of waves: interference and diffraction

Be a warm day I fancy. Specially in these black clothes feel it more. Black conducts, reflects (refracts is it?) the heat.

James Joyce: Ulysses

In the geometrical optics approximation we made the tacit assumption that rays intersect without interacting with each other. This would mean that rays meeting at a point, as at the image of a point source formed by an aberration-free lens, produce an infinitely small point image; this is known experimentally to be untrue. It would also mean that, if two beams of light overlap on a screen, the resultant light intensity (power density) is the sum of the intensities in the individual beams; experimentally this is sometimes true, sometimes false. In this chapter we examine these effects, which are, of course, examples of *diffraction* and *interference*.

3.1. Interference of two beams

We saw in Chapter 1 that nominally monochromatic light beams have very rapid random phase variations. Thus in order to see interference effects between two beams we must ensure that these phase variations are the same and in step in both beams. This is done by taking both beams from the same light source. It is simplest to think first about beams of plane waves intersecting at an angle θ. There were many classical experiments in which this was done in different ways. Figure 3.1 shows one way in which it might be done with modern equipment. If the region where the beams intersect is examined, e.g. by putting a white screen there or by scanning a small photodetector across it, straight dark and light bands perpendicular to the plane of the diagram are found, i.e. *interference fringes. Bright fringes are formed whenever the two waves are in phase.* The inset to Fig. 3.1 shows wavefronts from the two beams at a given instant and from this it can be seen that the spacing σ between the fringes is given by

$$\sigma = \lambda/\sin \theta, \tag{3.1}$$

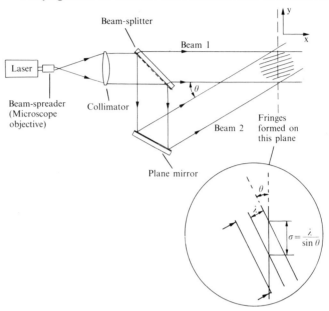

FIG. 3.1.   A convenient way to form interference fringes. Two beams of collimated light from the laser intersect at an angle θ, and fringes are formed where the beams cross. The beam-splitter is a plate of glass with a thin semi-transparent film of aluminium on one surface.

since it corresponds to the intersection of one wavefront of beam 1 by successive wavefronts of beam 2.

In slightly more detail, if we suppose the beams are of equal intensity we can represent their complex amplitudes by (Section 1.3),

$$\text{beam 1: } E \exp\{-2\pi i z/\lambda\},$$
$$\text{beam 2: } E \exp\{-2\pi i(z \cos \theta + y \sin \theta)/\lambda\}.$$

(3.2)

The total complex amplitude in the interference pattern across the plane $z = 0$, which we can take to be the plane of observation, is then

$$E\{1 + \exp(-2\pi i(y/\lambda) \sin \theta)\},$$

and the observed intensity is the squared modulus of this, i.e.

$$I(y) = 2E^2\left\{1 + \cos\left(\frac{2\pi}{\lambda} y \sin \theta\right)\right\}$$
$$= 4E^2 \cos^2\left(\frac{\pi}{\lambda} y \sin \theta\right),$$

(3.3)

This function, giving what are usually called *cos² fringes* or *two-beam fringes*, is plotted in Fig. 3.2. Since each time the argument of the cosine increases by π we go through a complete period of the fringes, we have verified eqn (3.1) above for the fringe spacing. The $\cos^2$ light-intensity distribution can be verified by photo-electric scanning, but visually the fringes appear to have much narrower dark regions than in the figure. This is a consequence of the

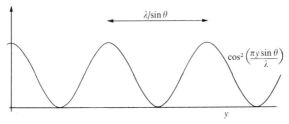

FIG. 3.2.   The light-intensity distribution across $\cos^2$ fringes.

very nonlinear response of the eye (Section 1.4), and it is often misleading in making a quick visual assessment of an interference or diffraction effect.

If the beams are of different amplitudes, and therefore of different intensities, the minima in the fringe system will not be zeros, i.e. there will not be maximum contrast or visibility in the fringes (see Problem 3.3).

Generally, we are mainly interested in the fringe spacing and contrast rather than in the details of the intensity variation in the fringes. The maxima occur where the two beams are in phase. We can generalize this immediately by noting, from Chapter 2, that points of equal phase occur where the optical path lengths from the source via the two interfering beams to the point in question are the same, or where they differ by a whole number of vacuum wavelengths. This is then applicable to interference in media of different refractive index. A good example is the oil film on water or, more generally, a layer of index n and thickness d, as in Fig. 3.3. Let a collimated (parallel) beam of wavelength λ meet the layer at an angle of incidence I. Some light is reflected at each surface, and there will be a path difference between corresponding wavefronts. This is indicated in the figure where Σ_1 and Σ_2 have originated from the same wavefront after a certain time. It is a simple exercise in the use of Snell's law (Chapter 2) to show that the optical path difference between the two beams is

$$2nd \cos I', \tag{3.4}$$

where I' is the angle of incidence *inside* the layer. Thus we should expect a bright fringe to be formed in the film whenever

$$2nd \cos I' = N\lambda \tag{3.5}$$

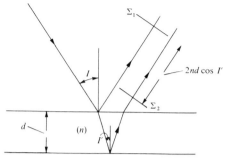

FIG. 3.3.   The difference of optical paths between beams reflected at the first and second surfaces of a layer of thickness d and refractive index n.

where N is an integer, the *order of interference*. In fact under certain conditions there is a change of phase of π of one or both of the beams on reflection and so the right-hand side of eqn (3.5) would be written $(N + \frac{1}{2})\lambda$ if the phase change occurred at one but not the other surface. The phase change on reflection is obtained from electromagnetic theory (see e.g. F. N. H. Robinson, *Electromagnetism*, OPS 1); here it is sufficient to note that for angles of incidence below the Brewster angle (Section 4.3) or for most practical purposes below 45° a phase change of π occurs when the light is incident on the interface from the medium of lower refractive index, e.g. reflection at a water surface in air, but not when the light is incident from the medium of higher index. Thus if the film were of oil of refractive index 1.45 floating on water ($n = 1.33$) there would be a phase change at the air–oil interface but not at the oil–water interface and $(N + \frac{1}{2})\lambda$ should be used in eqn (3.5). Again, if Newton's rings are formed by placing a convex lens surface on a plane surface (Fig. 3.4) the value of d is zero at the point of contact, the centre of the fringe pattern, and the centre will be *dark*.

Equation (3.5) shows how variations in the thickness d of an oil film are indicated by the shape of the interference fringes. Also in Newton's rings (Fig. 3.4) the successive fringe diameters indicate the variation of the thickness of the gap between the two glass surfaces. This equation also has other applications, as will be seen in the next section and in Chapter 6.

3.2. Interference with extended and polychromatic light sources

In discussing interference between the beams reflected from an approximately parallel layer in the previous section we assumed the incident light was monochromatic and collimated over a reasonable area of the layer, and this implied that it came from a single point source at a great distance. However, we normally see such effects under less stringent conditions: the source may cover a large extent, e.g. the sky for oil films or a sodium lamp for Newton's

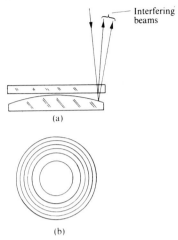

Interfering
beams

(a)

(b)

FIG. 3.4.   Newton's rings formed by interference between beams reflected at a curved interface. (a) The apparatus. (b) The spacing to scale of successive bright fringes. If the surfaces are in contact the central fringe is dark.

rings, and it may have a range of different wavelengths in it. We can understand this by referring again to eqn (3.5).

First, consider an extended source. Each source point will form its own interference fringes, and these fringe systems will be independent, i.e. the intensities will add. The angle of incidence I' will be different for different source points, so that the path difference (eqn (3.4)) at a given part of the layer will vary, and thus the fringe maxima will not coincide. However, if the thickness d is small enough quite a large variation in the angle of incidence is needed to change the path difference by, say, $\lambda/4$, so that the fringe systems will all more-or-less coincide, and fringes can be seen with an extended source. Problem 3.5 illustrates this. In Chapter 1 we spoke of the need to have *coherence* between light beams if they are to show interference effects. In the present case we see that the source has to be restricted in size (more exactly in angular subtense) to ensure that the beams reflected from the two surfaces of the layer are coherent.

If we now put eqn (3.5) in the form

$$\frac{2\,nd\cos I'}{\lambda} = N,\tag{3.5a}$$

we see that for given thickness and angle of incidence the order of interference N will vary with the wavelength. If N is non-integral it is interpreted as the number of wavelengths, possibly fractional, of path difference between the interfering beams. Thus if the source is polychromatic the fringe systems from the various wavelengths will again be displaced and will add in intensity to give a more-or-less uniform appearance, as in Fig. 3.5. However, again if the layer is thin enough, i.e. d is small enough, N in eqn (3.5a) will vary very little over a reasonable wavelength range, and fringes of good contrast will be obtained. Thus for coherence we have to restrict the wavelength range.

Two-beam interference effects are used in many devices. A few of these devices are discussed in Chapter 6, and others are described in classical texts (e.g. Ditchburn 1952; Longhurst 1973). From the above we can make a generalization which applies with appropriate modifications to all of these.

Two-beam interference effects can be obtained with polychromatic extended light sources. The contrast or visibility of the fringes depends on both the bandwidth (frequency or wavelength spread) and the angular extent of the source. Generally both of these must decrease with increasing optical path difference between the beams if the visibility or contrast is to be kept high.

In Chapter 6 we shall see applications of this principle to astronomy and spectroscopy.

FIG. 3.5.   Superimposition of intensities of fringes formed in polychromatic light.

3.3. Diffraction

In the previous section we started by discussing interference between collimated beams, as in Fig. 3.1. The beams were regarded as composed of plane waves of indefinitely great width—i.e. simply as described by, say, eqn (1.5)—with no restriction placed on the position vector which indicates the point in space at which we consider the wave disturbance. In fact the beams are limited in extent by the diameters of the collimator lenses in Fig. 3.1, and these certainly cannot be considered as indefinitely large. This restriction does not materially affect the description of two-beam interference; however when we examine the propagation of a single collimated beam from a point source we find that it does not propagate indefinitely with a sharply defined rim given by the edge of the lens; instead, the disturbance spreads out and becomes uneven near the edge in a complicated way. This is indicated in Fig. 3.6, which shows the light intensity observed in line with the edge at different distances. We now discuss this effect, known as diffraction.

We can simplify the discussion by considering first a collimated beam of large diameter which meets an opaque straight edge: experimentally the effect is very much as in Fig. 3.6, i.e. it does not much matter if the edge is curved or

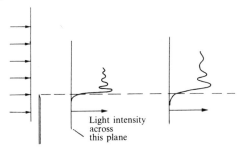

FIG. 3.6.   Diffraction at an edge. The graphs show the light intensity near the geometrical shadow at two distances from the edge in the ratio 1:4.

straight. Similar effects are found in the propagation of all waves; in sound and water waves it is easy to observe diffraction at an edge or any obstacle because the wavelength is large. Qualitatively we can see why diffraction occurs (i.e. why there is not a sharp shadow at all distances from the edge in Fig. 3.6) in terms of a basic physical principle that discontinuities do not occur in the wave representation. Thus the wave disturbance cannot stop abruptly at the line of the geometrical shadow, but must decay gradually. However, this does not give us a quantitative picture.

It can be shown (Chapter 10 of *Electromagnetism*, OPS 1) that in a uniform medium the electric and magnetic fields of electromagnetic radiation both obey a partial differential equation—the equation of wave motion. Thus for one component of the electric field vector, E_x say, we have, in a uniform dielectric medium

$$\nabla^2 E_x = \mu_r \mu_0 \varepsilon_r \varepsilon_0 \ddot{E}_x, \tag{3.6}$$

where μ_0 and ε_0 are the permeability and permittivity of vacuum and μ_r and ε_r are the relative permeability and the relative permittivity of the medium. For monochromatic (single-frequency)waves we can put $E_x = E \exp i\omega t$ in eqn (3.6), where E now becomes a complex amplitude (Chapter 1), and we obtain

$$\nabla^2 E + \omega^2 \mu_r \mu_0 \varepsilon_r \varepsilon_0 E = 0, \tag{3.7}$$

the time-independent equation of wave motion.

To solve the diffraction problem in complete generality for electromagnetic waves we should have to solve eqn (3.7) and five others like it for all components of **E** and **H**, using appropriate distributions of μ_r and ε_r and putting in suitable boundary conditions at the diffracting obstacles. This has been done for some simple cases, and the results have been verified experimentally by measurements with microwaves ($\lambda \sim 10$ mm). In the optical region we can greatly simplfy the problem by considering just one component of **E** (the scalar wave theory) and by using some approximations which are good for most regions of practical interest. Roughly, these regions are those at a large distance from the diffracting structures, where 'large' means many wavelengths and where the diffracting angles are small, i.e. less than, say, 0·1 rad. Figure 3.7 illustrates these regions for diffraction of a wave at an aperture in a screen.

We then have the following physical picture. In wave propagation, as the disturbance progresses through the medium, each point reached by the

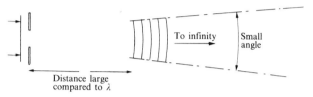

FIG. 3.7.   The region in which scalar diffraction theory can be safely applied.

disturbance becomes in turn a starting point for the disturbance which moves on from that point. This is easily pictured in the case of a transverse disturbance travelling along a stretched string or for water waves from an origin. For diffraction of a plane wave at an aperture in a screen we proceed as in Fig. 3.8. The wave travels from left to right and meets a plane opaque

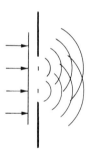

FIG. 3.8.   Huygens' secondary wavelets. Two secondary sources are indicated in the aperture.

screen with an aperture in it. To find out what happens on the right of the screen we assume that from each point in the plane of the screen secondary spherical wavelets spread out. The resultant disturbance is to be obtained by adding the complex amplitudes of all of these, *taking account of their relative phases*. Thus we consider interference between different elements of the same disturbance in the plane of the screen. If the screen were not there we should expect this curious procedure to give the same result as if the plane waves were simply carried on, and this is found to be so.

These ideas were roughly formulated by Huygens in the seventeenth century, refined by Fresnel early in the nineteenth century, and given a definite mathematical form by Kirchhoff about 80 years later. Huygens and Fresnel put their ideas in the form of the physical picture we have sketched but Kirchhoff obtained his result as a solution of eqn (3.7), with certain simplifying assumptions which restrict the applicability as outlined above. We give here a special case which is simpler than Kirchhoff's formulation but which applies to many problems of current interest.

We take a rectangular coordinate system as in Fig. 3.9 with the x, y-plane

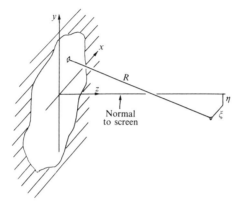

FIG. 3.9.   The diffraction integral. The distances ξ and η as shown are both negative.

as the plane of the screen, so that the arriving wavefronts are parallel to this plane, and we wish to determine the complex amplitude at a point P with coordinates (ξ, η, ζ). According to the above ideas, we assume that an element $dx\,dy$ of the wavefront in the aperture acts as a source of a secondary spherical wavelet of strength proportional to the area of the element. Thus the complex amplitude at P due to this element is

$$\frac{E_0}{\lambda} \cdot \frac{\exp(-ikR)}{R}\, dx\,dy,$$

where $k = 2\pi/\lambda$ as in Section 1.3. E_0 is a constant complex amplitude. The factor E_0/λ occurs naturally in the development according to Kirchhoff. Here we shall accept the factor as a way of keeping the equations dimensionally correct. The total effect at P due to all the incident wave which passes through the aperture is then

$$E_P = \frac{E_0}{\lambda} \int\!\!\int \frac{1}{R} \exp(-ikR)\, dx\,dy. \tag{3.8}$$

The integral is to be taken over the whole of the aperture and R is, of course, a function of x and y as different elements of the wavefront are taken. To express R in a manageable way, we have, from Pythagoras' theorem,

$$R^2 = (x - \xi)^2 + (y - \eta)^2 + \zeta^2$$
$$= x^2 + y^2 - 2(\xi x + \eta y) + \xi^2 + \eta^2 + \zeta^2.$$

Thus

$$R = \zeta \left\{ 1 - \frac{2(\xi x + \eta y)}{\zeta^2} + \frac{x^2 + y^2}{\zeta^2} + \frac{\xi^2 + \eta^2}{\zeta^2} \right\}^{\frac{1}{2}}$$

$$= \zeta - \frac{\xi x + \eta y}{\zeta} + \frac{x^2 + y^2}{2\zeta} + \frac{\xi^2 + \eta^2}{2\zeta} + \dots, \tag{3.9}$$

on expanding as far as the first term by the binomial theorem. As we should expect, R consists of a term $\zeta + (\xi^2 + \eta^2)/\zeta$ which is independent of x and y and which is large in terms of λ, together with some smaller terms. Whenever R increases or decreases by one wavelength (as a consequence of varying x and y in the integration) the exponent in eqn (3.8) changes by 2π, and the complex amplitude in the integrand goes through a complete cycle. Thus small changes in R are important in the exponential. On the other hand, since we are supposing R is large compared with λ, such changes can be ignored in the $1/R$ factor, and this can be written $1/\zeta$ and taken outside the integral. Since we are assuming small diffraction angles we need only consider linear terms in ξ/ζ and η/ζ in eqn (3.9). Eqn (3.8) then becomes

$$E_P = \frac{E_0}{\lambda \zeta} \exp(-ik\zeta) \times \int\int \exp\left\{ \frac{ik}{\zeta}(\xi x + \eta y) - \frac{ik}{2\zeta}(x^2 + y^2) \right\} dx\, dy \tag{3.10}$$

We now make an important simplifying assumption, that ζ is so large that the term in $x^2 + y^2$ can be neglected. This means that the maximum value of $(x^2 + y^2)/\zeta$ anywhere in the aperture is much less than λ. The physical implication will be seen in the next section. The factor $\exp(-ik\zeta)$ is a constant which will give a factor of unity when we take the squared modulus of E_P to get the light intensity, so it can be dropped, and we have as the final expression for the complex amplitude at P,

$$E_P = \frac{E_0}{\lambda \zeta} \int\int \exp\left\{ \frac{i2\pi}{\lambda \zeta}(\xi x + \eta y) \right\} dx\, dy. \tag{3.11}$$

We can apply this immediately to the simple but useful case of a square aperture of side a, as in Fig. 3.10. We take the origin at the centre of the square, and we let ζ be the z-coordinate of the plane in which we want to find the diffraction pattern. The double integration in eqn (3.11) splits into two factors,

$$E(\xi, \eta) = \frac{E_0}{\lambda\zeta} \int_{-\frac{1}{2}a}^{\frac{1}{2}a} \exp\left(\frac{i2\pi}{\lambda\zeta} \xi x\right) dx \int_{-\frac{1}{2}a}^{\frac{1}{2}a} \exp\left(\frac{i2\pi}{\lambda\zeta} \eta y\right) dy$$

$$= \frac{a^2 E_0}{\lambda\zeta} \left\{\frac{\sin (\pi a\xi/\lambda\zeta)}{\pi a\xi/\lambda\zeta}\right\} \cdot \left\{\frac{\sin (\pi a\eta/\lambda\zeta)}{\pi a\eta/\lambda\zeta}\right\}.$$

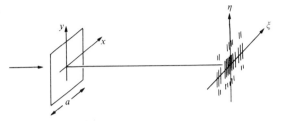

FIG. 3.10.   Far-field diffraction at a square aperture.

The notation sinc x is used for $(\sin \pi x)/\pi x$, so that the diffracted complex amplitude at (ξ, η) is

$$E(\xi, \eta) = \frac{a^2 E_0}{\lambda\zeta} \operatorname{sinc}\left(\frac{a\xi}{\lambda\zeta}\right) \operatorname{sinc}\left(\frac{a\eta}{\lambda\zeta}\right). \tag{3.12}$$

The light intensity $I(\xi, \eta)$ is the squared modulus of the complex amplitude (Section 1.3), so we have

$$I(\xi, \eta) = \frac{a^4 E_0^2}{\lambda^2\zeta^2} \operatorname{sinc}^2\left(\frac{a\xi}{\lambda\zeta}\right) \operatorname{sinc}^2\left(\frac{a\eta}{\lambda\zeta}\right). \tag{3.13}$$

The general form of this pattern, a central maximum of intensity with surrounding subsidiary maxima and minima, is indicated in Fig. 3.10 by the hatching (see also Problem 3.7). In describing such patterns we usually rescale the intensity by normalizing it to unity at the centre of the pattern, i.e. at $\xi = \eta = 0$. For eqn (3.13) this simply means omitting the factor $a^4 E_0^2/\lambda^2\zeta^2$; this is acceptable for most problems in physical optics, but the physical dimensions are lost. The factor $a^4 E_0^2/\lambda^2\zeta^2$ indicates that the central intensity is proportional to the fourth power of the linear dimensions of the aperture and inversely proportional to the square of the wavelength. These are general rules applying to all diffraction of this kind, where quadratic terms in the aperture are negligible. From the argument of the sinc function in eqn (3.13) the lateral scale, i.e. distances between successive maxima, varies inversely as the aperture size and directly as the wavelength, and these are again general rules.

Figure 3.11 shows this diffraction pattern quantitatively. The lines are contours of constant intensity in the pattern, sometimes called *isophots*, and

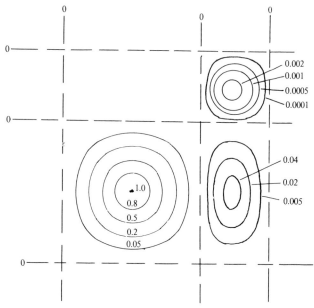

FIG. 3.11.   Contours of constant intensity in the diffraction pattern from a square aperture, in the far field; two of the side lobes nearest the central maximum are shown; the other six nearest side lobes are of the same shapes as these two, by symmetry.

the intensity scale is normalized to unity at the centre. The lateral scale can be obtained from the lines of zero intensity, since these correspond to values of π, 2π, 3π, ... in the argument of the $sinc$ functions in eqn (3.13).

We stress that the results given in this section are obtained after a succession of approximations from the original formulation of Kirchhoff, and that the latter itself is only an approximate way of solving the six partial differential equations, like eqn (3.7), obtained from Maxwell's equations. We justify the use of eqn (3.11) by the fact that experimentally it is verified to good accuracy in most ordinary situations.

3.4. Diffraction in the far field

Suppose that in Fig. 3.9 we observe the diffracting aperture from the point P at which the diffracted field is being found and suppose also that we could see the variations in complex amplitude in the aperture due to the variation in optical path length R from P to points (x, y) in the aperture (this could be done by, for example, a suitably arranged interferometer). The assumption we made that the term in $x^2 + y^2$ in eqn (3.10) is negligible means that ζ is taken so large that this phase variation over the aperture would be linear rather than quadratic, since it is in fact the exponent in eqn (3.10). The distance ζ from the diffracting aperture then satisfies

$$\zeta \gg a^2/\lambda, \tag{3.14}$$

and we have *far-field* diffraction or *Fraunhofer* diffraction. If on the other hand the quadratic term is not negligible then

$$\zeta < a^2/\lambda, \tag{3.15}$$

and we have *near-field* or *Fresnel* diffraction.

We shall consider first far-field diffraction. In the limit as ζ gets very large we can put $\xi/\zeta = u, \eta/\zeta = v$, where u and v are angular coordinates, and we speak of light diffracted into a direction (u, v). Eqn (3.13) is then written in normalized form

$$I(u, v) = \text{sinc}^2\left(\frac{au}{\lambda}\right)\text{sinc}^2\left(\frac{av}{\lambda}\right) \tag{3.13a}$$

We do not actually have to go to a distance given by eqn (3.14) to observe far-field diffraction. From Section 2.3 parallel rays in a direction with components (u, v) on one side of a lens come to a focus at a point with coordinates (fu, fv) on the focal plane of the lens. Thus an objective (collimator in reverse) can be used to bring the far field to a convenient distance, as in Fig. 3.12.

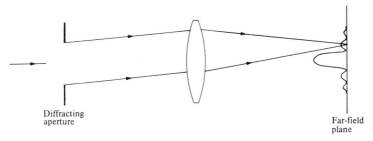

Diffracting aperture

Far-field plane

FIG. 3.12.   Bringing the far field to a convenient place for observation. The far-field diffraction pattern of the aperture is formed at the focal plane of the lens. (If in addition we want the phase relationships between different parts of the far-field pattern to represent correctly the Fourier transform of the aperture we must also put the aperture at the front focal plane of the lens.)

We can re-write eqn (3.11) normalized to have unity intensity at the centre of the far-field, i.e. in the direction of the incident wave, given by $\xi = \eta = 0$, and we can also write the far-field coordinates as the diffraction angles (u, v). We then have for the normalized complex amplitude at (u, v),

$$E(u, v) = \frac{1}{A}\iint \exp\left\{\frac{i2\pi}{\lambda}(ux + vy)\right\}\,\mathrm{d}x\,\mathrm{d}y, \tag{3.11a}$$

where A is the area of the diffracting aperture and, as before, the integration is over the aperture. Next we specify the diffracting aperture by means of a function $F(x, y)$, which is put as a factor inside the integral and which is defined as equal to unity for (x, y) inside the pupil and zero outside. This enables us to formally extend the limits of integration to infinity and eqn (3.11a) becomes

$$E(u, v) = \frac{1}{A} \int\limits_{-\infty}^{\infty}\!\!\int F(x, y) \exp\left\{\frac{\mathrm{i}2\pi}{\lambda} (ux + vy)\right\} \mathrm{d}x\, \mathrm{d}y. \qquad (3.11b)$$

The introduction of the function $F(x, y)$ is more than a mere formal device. It need not be only a binary function (i.e. taking only values 0 or 1); it can be modified to give the effect of a screen across the aperture which absorbs some light or which has a phase-changing effect. Both these devices are useful; if there is absorption the normalizing factor strictly has to be interpreted as $(\iint F(x, y)\, \mathrm{d}x\, \mathrm{d}y)^{-1}$.

Equation (3.11a) can be interpreted as a Fourier transform relationship (see Appendix). We see that if eqn (3.11b) is rewritten in the form

$$f(s, t) = \int\limits_{-\infty}^{\infty}\!\!\int \frac{F(x, y)}{A} \exp\left\{\mathrm{i}2\pi(sx + ty)\right\} \mathrm{d}x\, \mathrm{d}y \qquad (3.11c)$$

where we have put $s = u/\lambda$, $t = v/\lambda$ as new variables and where $E(\lambda s, \lambda t) \equiv f(s, t)$, we can say that the complex amplitude in the far-field diffraction pattern is the (inverse) Fourier transform of the complex amplitude in the diffracting aperture, to a suitably chosen scale.†

We can see from eqn (3.11a) that the size of the far-field pattern is proportional to the wavelength and inversely proportional to the linear scale of the diffracting aperture; this is, from the general properties of Fourier transforms, (Appendix) true for any shape of aperture, in agreement with the rule stated in the previous section.

We meet far-field diffraction in many different physical situations. A microwave paraboloid antenna has a characteristic angular distribution of radiated power which is the far-field diffraction pattern of the 'aperture' formed by the rim of the paraboloidal reflector. If we could avoid certain distorting effects due to atmospheric turbulence, the image of a star formed by a telescope would be the far-field diffraction pattern of the aperture of the telescope mirror. The Laue X-ray diffraction pattern from a single crystal is

---

† The question of whether it is the inverse or direct transform is not physically significant. The sign of the imaginary unit (i.e. $\pm\mathrm{i}$) in the complex amplitude can be chosen arbitrarily in the first instance, since we are only dealing with the real part of the complex amplitude.

the far-field diffraction pattern of a regular lattice of diffracting points, i.e. it is the Fourier transform of a periodic array of delta functions.

The far-field pattern for a circular aperture must be radially symmetrical about the axis of the aperture. Taking only one angular coordinate u, which is the (small) angle between the axis and the direction in the far-field in which we are interested, it is found that the complex amplitude in the far-field is in normalized form

$$E(u) = \frac{2J_1(2\pi au/\lambda)}{2\pi au/\lambda}, \qquad (3.16)$$

where $J_1(z)$ is the Bessel function of the first kind and first order. This function, which is available from many books of tables, behaves like an attenuated sine wave.† The intensity in the diffraction pattern is plotted in Fig. 3.13 with logarithmic ordinate scale. It is known as the *Airy diffraction pattern*, after the astronomer G. B. Airy, who calculated it as the theoretical form of a star image.

We can observe far-field diffraction effects in the optical region most easily with a laser as light source. It is necessary that all parts of the wavefront in the diffracting aperture should be able to interfere with each other, i.e. they must be coherent with each other in the sense explained in Section 1.5. The light from a helium-neon laser in correct adjustment behaves as if it came from a single point source, and it is therefore coherent over the whole wavefront. Figure 3.14 shows a typical arrangement of apparatus for producing far-field diffraction patterns; diffracting screens of different shapes are placed in the collimated beam. Many beautiful examples of far-field diffraction patterns are given elsewhere (e.g. Lipson 1972).

It is often useful to estimate the general features of a far-field diffraction pattern without a detailed calculation. It is usually true that if the diffracting aperture is simple in shape and has no blanked-off area in the middle then the maximum intensity in the far-field is in the direction of the incident wavefront. If we go away from this direction by an angle λ/d, where d is a distance of the order of magnitude of the width of the aperture, this will be roughly the direction of the first minimum. Thus the angular half-width of the central maximum, i.e. the full width of the pattern at half maximum intensity, is of order λ/d. For example, a laser beam about 1 mm in diameter (i.e. as it comes from the laser) will spread by diffraction, even if nominally collimated, over an angle of about 1 mrad, but if the beam is first spread out by means of a beam-expander (as in Fig. 3.15) to, say, 20 mm, it will only spread at about 0.05 mrad.

---

† A comprehensive collection is *Handbook of mathematical functions with formulas, graphs and mathematical tables* by M. Abramovitz and I. Stegun (National Bureau of Standards, Washington, DC, 1965).

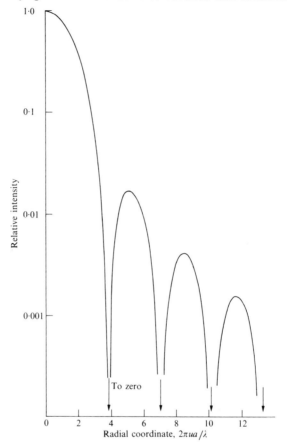

FIG. 3.13.   The Airy pattern, or light-intensity distribution in the image of a point source of monochromatic light formed by a system with a circular aperture. The function plotted is the squared modulus of the amplitude, as in eqn (3.16).

3.5. Diffraction in the near field

On account of the restriction mentioned in Section 3.3 we cannot use our simple formulation to examine diffraction effects right up to the diffracting aperture, but we can ask what happens in the region nearer the aperture than the far-field, where the term in $x^2 + y^2$ in the integral of eqn (3.10) cannot be neglected. This is the Fresnel or near-field region. We take as a simple example a very long slit of width a, so that the other dimension can be neglected and we have only a single integral to deal with. We shall calculate

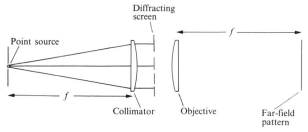

FIG. 3.14.   A diffractometer. The far-field pattern of the screen is formed at the focal plane of the objective. If the complex amplitude in the far-field pattern is to have the correct phase distribution according to eqn (3.11c) the screen should be at a distance f from the objective, but this is not necessary if only the intensity is to be observed.

FIG. 3.15.   A beam-expander. This is an afocal system, i.e. it forms an image at infinity of an object at infinity. The incoming beam is expanded in the ratio of the focal lengths of the two lenses.

the complex amplitude at a distance ζ from the plane of the slit and we take a coordinate ξ as in Fig. 3.16. Then from eqn (3.10) we have to evaluate

$$E_P = \int_{-\frac{1}{2}a}^{\frac{1}{2}a} \exp\left\{\frac{ik}{\zeta}(\xi x - \tfrac{1}{2}x^2)\right\} dx$$

$$= e^{\frac{i\pi\xi^2}{\lambda\zeta}} \int_{-\frac{1}{2}a}^{\frac{1}{2}a} \exp\left\{\frac{-i\pi}{\lambda\zeta}(x - \xi)^2\right\} dx. \qquad (3.17)$$

We can drop the factor outside the integral since it will give unity on taking the squared modulus to get the intensity, and eqn (3.17) can then be expressed in terms of Fresnel integrals; these are defined as follows

$$C(z) = \int_{0}^{z} \cos\frac{\pi}{2}t^2 \, dt$$

$$S(z) = \int_{0}^{z} \sin\frac{\pi}{2}t^2 \, dt. \qquad (3.18)$$

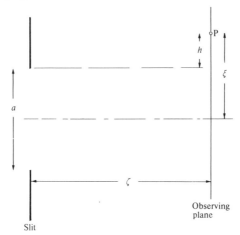

FIG. 3.16.   Near-field diffraction from a slit.

After some simple changes of variable (eqn (3.17) gives

$$\sqrt{\left(\frac{2}{\lambda\zeta}\right)}E_P = C\left\{\sqrt{\left(\frac{2}{\lambda\zeta}\right)}(\tfrac{1}{2}a + \xi)\right\} \pm C\left\{\sqrt{\left(\frac{2}{\lambda\zeta}\right)}|\tfrac{1}{2}a - \xi|\right\}.$$
$$+ i\left[S\left\{\left(\frac{2}{\lambda\zeta}\right)(\tfrac{1}{2}a + \xi)\right\} \pm S\left\{\sqrt{\left(\frac{2}{\lambda\zeta}\right)}|\tfrac{1}{2}a - \xi|\right\}\right]. \tag{3.19}$$

The positive signs are taken if ξ is less than $\tfrac{1}{2}a$, the negative if ξ is greater than $\tfrac{1}{2}a$.

We can use this result to get the diffraction pattern across the shadow boundary of a straight edge, by putting $\xi - \tfrac{1}{2}a = h$, say, so that h is the distance from the boundary, and letting a tend to infinity. From tables of the Fresnel integrals $C(\infty) = S(\infty) = 0.5$, so that the complex amplitude across the shadow edge is

$$\tfrac{1}{2} \pm C\left\{\sqrt{\left(\frac{2}{\lambda\zeta}\right)}|h|\right\} + i\left[\tfrac{1}{2} \pm S\left\{\sqrt{\left(\frac{2}{\lambda\zeta}\right)}|h|\right\}\right] \tag{3.20}$$

where again the positive sign is to be taken if P is on the bright side of the shadow and the negative if it is on the dark side.

Figure 3.17 shows the light intensity in the edge diffraction pattern, i.e. the squared modulus of eqn (3.20). The scale of the pattern is proportional to the distance from the diffracting edge and to the wavelength. The distance of a fringe on the bright side of the shadow from the geometrical shadow line is approximately proportional to the number of the fringe, as with Newton's rings.

Since the width a of the slit has been increased to infinity to give the

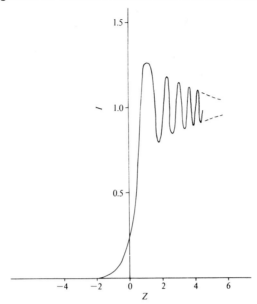

FIG. 3.17.   The near-field diffraction pattern from a straight-edge. The ordinate is
the light intensity, given by the squared modulus of the expression (3.20); the abscissa z
is the argument $\sqrt{(2/\lambda\zeta)}|h|$ of the Fresnel integrals. The geometrical optics shadow edge
is at $z = 0$.

diffraction pattern from a single edge it can be seen that no matter how large ζ
is taken the terms in $x^2 + y^2$ cannot be neglected in the diffraction integral, i.e.
it is a near-field pattern at any distance. Thus in observing the occultation of
a star by the Moon or by a planet the shadow sweeping over the surface of the
earth has a light intensity distribution similar to Fig. 3.17 to a suitable scale,
but with due allowance for the spread of wavelengths in the light of the star.

3.6. Interference, diffraction, and the photon picture

Interference and diffraction are now thought of as essentially wave phenomena, although there have been attempts (e.g. by Newton) to explain
interference using a particle model. Yet the dual nature of light—wave and
particulate—is well established, and the apparent contradiction at an
elementary level between these aspects is perhaps more striking than for any
other particle-cum-wave. The contradiction is sometimes handled as in
Section 1.6 simply by saying that we have to use the wave representation for
some purposes and the particle representation for others, but it is possible to
enlarge on this approach.

Suppose we have an interferometer, such as in Young's experiment (Section 6.1), and we wish to discuss its properties in terms of photons rather than waves. Then according to the methods of quantum mechanics we have to consider the passage of a single photon at a time through the apparatus. Indeed, with an ordinary light source and an apparatus of reasonable size, it can easily be shown that it is unlikely that more than a single photon will be in transit through the apparatus at any given time. Thus we imagine a detector, in the plane where the fringes are to be formed, which builds up the fringe pattern as dots, one for each arriving photon in the position where it activates the detector. This experiment was first done by G. I. Taylor using photographic plates, and it has since been repeated in many different ways. The results always show that the photons appear at first to be arriving at random positions, but gradually, as more photons arrive, the classical interference pattern as predicted by wave theory is built up.

We explain this by saying that the photon does not follow a definite path through the apparatus, but that it can follow any of several different paths. Clearly there should be a high probability for the photon to follow a path terminating near the position of an interference fringe maximum, and a low probability for it to arrive near a minimum. To calculate these probabilities we should have to solve Schrödinger's equation, putting in as boundary and initial conditions the shapes and sizes of the slits, etc. in the apparatus and the momentum and point of origin of the photons. If we were to do this we should find that we arrived eventually at the same form for the fringe system as that obtained by elementary wave theory. Thus we find the probabilities for the arrival of photons giving the same interference or diffraction patterns as the intensity distributions of interfering or diffracted waves.

In the above hypothetical calculation, although all the details of the apparatus have to go into the equations, all we get out at the end is a probability density for photons to be detected in a certain region. Thus we have by-passed the tantalizing question, which of the two slits in the interferometer did a certain photon go through? According to the rules of quantum mechanics we are not permitted to ask this question. We may only ask questions whose answers can be determined experimentally without upsetting the outcome of the main experiment. If we were to attempt to find out which slit the photon went through, we should have to annihilate it in the process of detecting it at one or other slit. Thus it would then not take part in forming the interference pattern which was the original object of the experiment.

We can also interpret coherence using the photon concept. If the photons traversing the apparatus all have nearly the same probability-density distributions in the interference region then the apparatus is illuminated coherently, since all photons will be contributing to the same fringe pattern. In practice, this means that the photons must all have nearly the same momentum and must come from nearly the same source point. This then corresponds to the classical condition, since momentum includes frequency and direction.

The above explanation is only a sketchy attempt to explain a topic which is in detail very complicated. However, much the same argument applies to interference and diffraction of all elementary particles. From a particle viewpoint we are dealing with varying probabilities of arrival at different points, and the wave picture provides, in effect, a convenient way of calculating these probabilities. Then, after calculating the probabilities by using wave theory, we can call them intensities if we are dealing with a large enough flux of particles.

Problems

3.1. Two beams of radio waves of frequency 3 MHz intersect at an angle of 10°. What is the interference-fringe spacing?

3.2. How many fringes are formed per millimetre if light beams of wavelength 632.8 nm intersect at 5°?

3.3. Two beams interfere at an angle θ. If the complex amplitudes are in the ratio 2:1, show that the intensity in the fringe system has the form

$$E^2 \left\{ 5 + 4 \cos \left(\frac{2\pi}{\lambda} y \sin \theta \right) \right\},$$

and plot this function.

3.4. Two glass plates are nearly in contact and make a small angle θ with each other. Show that the fringes produced by interference in the air film have a spacing equal to $\lambda/2\theta$ if the light is incident normally.

3.5. A monochromatic source of wavelength 546 mm is 25 mm in diameter and is placed 500 mm above an air film between two glass plates. Show that the air film can be about 0.2 mm thick before the fringes begin to lose visibility. (*Hint.* The range of angles of incidence is from 0° to $\theta = \arctan 12.5/500$; the range of path differences is from $2d$ to $2d \cos \theta$, and this should be less than $\lambda/4$.)

3.6. An air film is 100 μm thick and fringes are to be formed in it from a polychromatic source of mean wavelength 550 nm. Approximately what wavelength range can be used? (Choose a range from λ_1 to λ_2 such that N does not vary by more than 1/4.)

3.7. Plot the functions sinc x and sinc² x for values of x up the the third zero. Tabulate the values of the subsidiary maxima.

3.8. An aperture 5 mm in diameter diffracts light of wavelength 0.5 μm. How far away must a screen be placed to show the far-field diffraction pattern?

3.9. A teaching laboratory has a 2 m long optical bench. Suggest a suitable aperture size for demonstrating far-field diffraction with a helium-neon laser.

3.10. Plot the amplitude and intensity distribution in the Airy pattern. Find by numerical or graphical approximation the intensity in the first bright ring, the radius of the first dark ring, and the radius at which the intensity is half the central maximum.

3.11. A beam from a ruby laser (694 nm wavelength) is to be used in measuring variations in the distance of the moon by timing its return from mirror systems arranged on the moon. If the beam is expanded to 1 m diameter and collimated, estimate its size at the moon. (Moon's distance $\sim 3.8 \times 10^5$ km.)

4. Polarization

4.1. Everyday aspects

Most of us have observed polarized light, through 'Polaroid' sunglasses. What we are seeing arises because light can have asymmetry about the direction of propagation; the appearances change when the glasses are rotated. Thus in the wave representation of light the disturbance cannot be along the direction of travel, as it is in sound waves. Similarly, television aerials also have directionality, indicating that they are sensitive to electromagnetic waves in which the disturbance is perpendicular to the direction of propagation.

In radio waves it is the electric field which is usually of importance in detection, and thus the direction of the electric field is the polarization direction. We define the *plane of polarization* as a plane containing the direction of propagation and the electric field vector at any point in the radiation field. In the optical region the detection processes are associated almost entirely with the electric field (although in the photon picture this is not quite so easily formulated), so that the electric vector again defines the plane of polarization.†

Radio waves are polarized by emission from a transmitting aerial of the right shape (see *Electromagnetism*, OPS 1), but light from everyday sources is sometimes almost completely unpolarized. Thus in the model (Section 1.5) of a light beam with randomly varying phases and amplitudes we have to add that the state of polarization, which we shall define below, is also changing rapidly and randomly in time. This is explained by saying that the states of polarization of wave-trains emitted by different atoms have no fixed relationships. However, if we require polarized light, it is very easy to obtain from an unpolarized beam, by passing the beam through a polarizer.‡

---

† The plane of polarization was originally chosen by arbitrary convention to be that containing, as we now know, the magnetic field vector, and this definition can be found in texts published up to about 50 years ago.

‡ In the rest of this chapter we shall use 'polarizer' to mean a piece of 'Polaroid' or any other device, such as a Nicol prism, which can produce polarized light.

4.2. Kinds of polarized light

To find out what a polarizer does we must first describe polarized light. Suppose we have a beam of collimated light travelling in the z-direction, as in Fig. 4.1. To polarize it we put a polarizer in the beam with its polarizing direction parallel to the y-axis,† and we then have the beam polarized with the

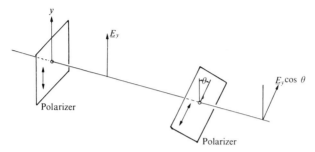

FIG. 4.1. The light transmitted by two polarizers at an angle θ.

electric vector in the y-direction. If we follow this with a second polarizer with its axis at right-angles to that of the first then no light is transmitted. If the second polarizer has its axis at some other angle θ we resolve the incident electric field E_y into components $E_y \cos \theta$ and $E_y \sin \theta$ parallel and perpendicular to the new direction, and only $E_y \cos \theta$ is transmitted. Thus we should expect the transmitted light *intensity* to vary as $\cos^2\theta$. Experimentally this is found to be so, and this is confirmation of our model of a polarizer and of polarized light.

The light produced by a polarizer as described above is said to be plane-polarized, because the electric vector remains parallel to one plane—that which contains the direction of propagation and the polarizing direction of the polarizer. We can propose other kinds of polarization, as follows. Light plane-polarized in the y-direction has an electric field of the form

$$\underline{E}_1(t, \underline{r}) = (0, E_y, 0), \tag{4.1}$$

where $E_y = E_2 \cos (\omega t - kz)$ and the notation (A_x, A_y, A_x) denotes the three components of the vector $\mathbf{A}$. These equations are merely special forms of eqn (1.5) in which the real component only is taken. We can now suppose added

---

† To find the polarizing direction, we use the polarizer as in sunglasses, i.e. we look through it at a horizontal smooth shiny surface and turn it so that the reflection from the surface is minimized. The polarizing direction of the polarizer is then vertical, as will be seen in Section 4.3.

to this a coherent beam travelling in the same direction but polarized at right-angles and with different phase and amplitude,

$$\mathbf{E}_2(t, \mathbf{r}) = (E_x, 0, 0),$$
$$E_x = E_1 \cos{(\omega t - kz + \varepsilon)} \tag{4.2}$$

(An experiment in which such an addition occurs is described in Section 4.4.) To examine the result we take $z = 0$ and we suppose first that there is no phase difference, i.e. $\varepsilon = 0$. Then at all times the resultant electric field makes a constant angle $\arctan{(E_2/E_1)}$ with the x-axis, and so we have plane-polarized light, but with the plane of polarization in the direction $\arctan{(E_2/E_1)}$. However if there is a phase difference, the direction of the resultant field will change with time; e.g. if $\varepsilon = \pi/2$ we have

$$E_x = -E_1 \sin\omega t, \tag{4.3}$$
$$E_y = E_2 \cos \omega t,$$

and so the tip of a vector representing the electric field traces out an ellipse (as in Fig. 4.2) with angular frequency ω. We then have *elliptically polarized light*.

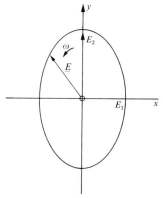

FIG. 4.2.   The electric field vector in elliptically polarized light. The vector rotates, tracing out the elipse, with angular frequency ω.

In order to make the distinction clear, we have to use 'plane-polarized' to describe the effect of an ordinary polarizer. The most general effect of adding two coherent plane-polarized beams with an arbitrary fixed phase difference between them is to produce elliptic polarization. If in eqn (4.3) $E_1 = E_2$ and $\varepsilon = \pi/2$, we have a special case—*circular polarization*—since the ellipse given by the equation becomes a circle.

In plane-, elliptic-, and circular-polarized light the phase difference between two orthogonal components of the electric field, resolved along any chosen axes, is constant in time. Unpolarized light, sometimes called 'natural light', can now be described in a more general way than in Section 1.5, as light in which the state of polarization, in general elliptic, changes rapidly and randomly in time. We can also have light beams which are mixtures of unpolarized and polarized (plane, elliptic, circular) light. These are said to be partially polarized. Most light from everyday sources is partially polarized, e.g. sky light, sunlight, light from metal filament lamps, and light reflected from smooth surfaces, but sometimes the polarized component is a small proportion of the total intensity.

In describing elliptic polarization a sign convention is necessary for the direction of rotation. The convention is that the rotation is clockwise looking towards the source for *right-hand* elliptic or circular polarization.

4.3. Production of polarized light

The reflection factor, or ratio of reflected to incident light intensity, for a smooth interface between transparent media of different refractive indices can be calculated for electromagnetic waves (see e.g. Chapter 10 of *Electromagnetism*, OPS 1). Let the light have a angle of incidence θ_1 from a medium of index n_1, as in Fig. 4.3. If the incident light is plane-polarized with the electric vector parallel to the plane of incidence, as in the figure, the reflection factor for light intensity is

$$R_{\mathrm{p}} = \left(\frac{n_2/\cos \theta_2 - n_1/\cos \theta_1}{n_2/\cos \theta_2 + n_1/\cos \theta_1} \right)^2, \tag{4.4}$$

and for the other polarization, with electric vector perpendicular to the plane of incidence, the reflection factor is

$$R_{\mathrm{s}} = \left(\frac{n_2 \cos \theta_2 - n_1 \cos \theta_1}{n_2 \cos \theta_2 + n_1 \cos \theta_1} \right)^2. \tag{4.5}$$

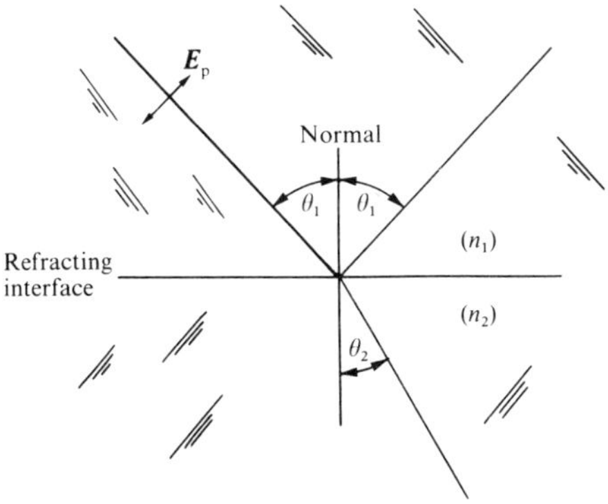

FIG. 4.3.   Reflection and refraction of p-polarized light at a dielectric interface.

62   Polarization

The subscripts p and s, which are almost universally used, stand for *parallel* and *senkrecht* (German for parallel and perpendicular). The ratios R_p and R_s are plotted in Fig. 4.4 for reflection at a glass–air interface, i.e. taking $n_1 = 1, n_2 = 1.5$. It can be seen that R_p is zero at an angle of incidence of about 57°, so that at this angle the reflected light will be completely plane-polarized perpendicular to the plane of incidence. It is easily

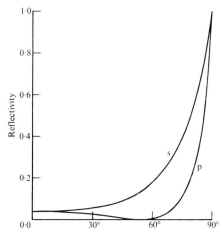

FIG. 4.4.   Reflection factor of a glass surface for p- and s-polarizations as a function of angle of incidence.

shown that there is always an angle of incidence which gives complete polarization. This is called the *Brewster angle*, and from eqn (4.4) it is given by

$$\tan \theta_B = n_2/n_1. \tag{4.6}$$

This result accounts for the action of polarizing sunglasses in reducing glare from horizontal smooth surfaces. Even if the reflecting surface is not on a transparent medium there is still considerable polarization at oblique incidence, so that it is easy to find approximately the plane of polarization of a polarizer by looking through it at obliquely reflected light.

Light is often partly polarized by scattering as well as by reflection. Thus blue sky light is sunlight scattered by air molecules, and it is strongly polarized for directions of scatter at large angles to the incident beam. This is easily seen through polarizing sunglasses.

Polarized light can also be produced by transmission through an *anisotropic* medium, i.e. a medium in which the optical properties (and many other physical properties also) vary with direction. All crystals except those of cubic structure are anisotropic and this considerably complicates their optical properties. The detailed mathematical description may be found in more advanced texts (e.g. Ditchburn 1976), where it is shown that for a given direction in a crystal two distinct plane waves can in general be propagated; they have different speeds and are polarized at right angles to each other. Also a plane wave refracted into such a crystal in general separates into two waves with different velocities and again orthogonally polarized. Another way to view this is in terms of propagation from a point source inside the crystal. For an isotropic medium the speed would be the same in all directions and the wavefronts would be spherical. For a crystal the surface which the disturbance would reach in a given time is called the *ray surface* (perhaps a rather confusing name in the present context) and it is found to be a double surface or surface of two sheets, corresponding to two oppositely polarized wavefronts travelling at different velocities.

Consider, for example, calcite (Iceland spar), which is crystalline $CaCO_3$. We suppose that by some means it is possible to produce a point source of light inside the crystal, as in Fig. 4.5. There are two parts to the propagating

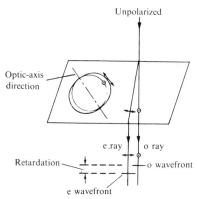

FIG. 4.5. Double refraction or birefringence in a crystal of calcite ($CaCO_3$). The angle between the o and e rays is exaggerated. The circles indicate that the electric field is perpendicular to the plane of the diagram.

disturbance, one a spherical wavefront (the o disturbance, wavefront, or beam) and the other an oblate ellipsoid of revolution (the e disturbance). The two systems propagate independently accordingly to Huygens' principle, and they are polarized at right-angles as indicated. Thus if a collimated beam is incident normally on the crystal, the o disturbance will pass through undeviated, but if we propagate the e disturbance as in Fig. 2.3), by Huygens' principle we find it is deviated as shown.† There is one direction in the crystal, the direction of the *optic axis*, such that both disturbances travel in the same

† The o and e rays were originally called the ordinary and extraordinary rays on account of this behaviour.

direction and with the same velocity. From the figure this is clearly the axis of revolution of the ellipsoid.

In most crystals the ray surface is more complicated than as described above, and it is better to consider propagation of plane waves rather than light from a fictitious point source in the crystal. In general, corresponding to any unpolarized plane wave incident on the crystal from air, there are two plane-polarized waves propagated inside the crystal in different directions and with different speeds. Two beams plane-polarized at right-angles and parallel in direction emerge from the crystal, as in Fig. 4.5. The lateral displacement between the beams is used in some polarizing devices, but a more important effect is a relative retardation or optical path difference between the two emergent wavefronts. The retardation is used to produce elliptic or circular polarization, as in Fig. 4.6. Plane-polarized light, with its

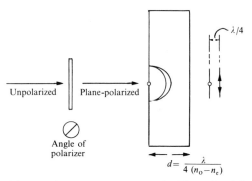

FIG. 4.6.   A calcite crystal as a quarter-wave retarder. The optic axis is parallel to the crystal surface (in the plane of the diagram), and the thickness is chosen to give $\lambda/4$ path difference through the crystal between the o and e rays. If the incident beam is polarized the two emergent beams are coherent, and they produce circularly polarized light for the arrangement shown.

plane of polarization at 45° to the optic axis of the crystal, is resolved in the crystal into two orthogonal components, with electric vectors respectively perpendicular and parallel to the optic axis. The first becomes the o beam and the second the e beam, and they propagate in the same direction (since the optic axis is parallel to the surface) but with different velocities. If the relative retardation when they emerge is $\lambda/4$ or any odd multiple of $\lambda/4$ they combine, as in Section 4.2, to produce circular polarization, and the crystal plate is said to be a quarter-wave plate. Other thicknesses (except those giving $\lambda/2$ or a multiple of $\lambda/2$ retardation) would give elliptic-polarization. It is customary to define refractive indices n_o and n_e corresponding to the velocities of propagation of the o and e beams in a direction perpendicular to the optic axis. The retardation as an optical path length is then $(n_e - n_o)d$, where d is the thickness of the crystal plate.

Isotropic materials, e.g. glasses, become *double refracting* or *birefringent*, like crystals, when under mechanical stress or in static electric or magnetic fields. Materials such as stretched plastic sheet (almost any kind) are found to be birefringent on inspection between polarizers. These materials contain long polymer molecules which acquire a partial common alignment from the stretching, and thus they are similar to crystals. If a chromophore is added to the polymer it may absorb one type of polarization and transmit the other; this is the principle of the commonest kind of 'Polaroid', which is a poly(vinyl alcohol)-iodine complex.

The effects of electric and magnetic fields in producing birefringence have applications in modern optics, e.g. in modulating the intensity in a light beam according to an electrical signal. There is a second group of effects in which a medium *rotates* the plane of polarization of incident plane-polarized light. This effect is intrinsic i.e., does not depend on electric or magnetic fields in certain solutions of molecules which form stereoisomers, i.e. the molecule and its mirror image cannot be superimposed. A simple example is lactic acid, $(CH_3)CH(OH) . CO_2H$, in which the central carbon atom is bonded to four different groups—methyl, hydrogen, hydroxyl, and carboxyl—so that this structure cannot match its mirror image. Many organic compounds have this property, including some sugars, and an important method, saccharimetry, of estimating sugar concentration is based on it. Rotation of the plane of polarization can be induced in almost all materials by a magnetic field. For transparent materials this rotation is called the *Faraday effect*. This effect has been used, for example, for estimating magnetic fields in space and for measuring very large direct currents (by estimating the surrounding magnetic field). Magnetic rotation also occurs on reflection from metal surfaces, when it is called the *Kerr effect*. This effect is used in studying the microstructure of magnetic alloys with a polarizing microscope.

Mathematical formulations and a detailed treatment of crystal optics and of electro- and magnetooptical effects are given elsewhere (e.g. Born and Wolf 1965).

4.4. Polarization and interference

According to the discussion in Section 4.1 we must include the state of polarization in any precise discussion of coherence. If the interference experiment as in Fig. 3.1 is carried out with an unpolarized source, we find interference fringes as expected. However, if we polarize each interfering beam separately in orthogonal directions there is no interference, and we cannot produce interference by then rotating one of the planes of polarization to agree with the other by means of, for example, a half-wave plate in one beam (see Problem 4.6). We interpret this to mean that unpolarized light is to be regarded as the sum of two plane-polarized and *mutually incoherent* components which are orthogonal, i.e. their planes of polarization are at

66   Polarization

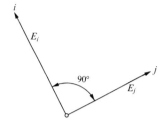

FIG. 4.7. Orthogonal plane-polarized components. They are mutually incoherent if selected from an initially unpolarized beam.

right-angles to each other. Thus in Fig. 4.7 we are looking along the unpolarized beam, and we resolve the electric field into components along the two orthogonal directions i and j. Then the experiment shows that the fluctuations in amplitude and phase which occur in E_i are entirely uncorrelated with those in E_j. From this it follows that a beam with an infinitely long coherence length, i.e. a 'perfect' laser beam, would be completely polarized, although not necessarily plane-polarized.

The above result is sometimes put in the form 'beams polarized at right-angles cannot interfere', but this is misleading. If the initial beam in the interference experiment described on p. 38 had been polarized and then the split beams polarized as above with polarizers at $\pm 45°$ to the original direction of polarization, we should not see interference fringes, but the beams would be coherent and they would interfere. In fact this is just the experiment described in Fig. 4.6, where the coherent beams which are orthogonally polarized interfere to produce elliptically polarized light. If a single component is selected at the end by means of a polarizer parallel to the original direction, then interference fringes can be seen.

To summarize this argument, orthogonally polarized beams do not interfere if they are derived from two orthogonal components of unpolarized light. They do interfere, producing in general elliptic polarization, if they are derived from a single polarized beam.

Problems

4.1. A polarizer is placed in a beam of plane-polarized light of unit intensity, with its polarizing direction at an angle θ to the electric vector. Find (a) the modulus of the amplitude and (b) the intensity of the transmitted beam for $\theta = 10°$, $45°$, $88°$.
4.2. Elliptically polarized light has its axes along the x- and y-directions, and the relative field strengths are in the ratio $1:2$. Calculate the relative intensity transmitted by a polarizer with its plane (a) parallel to the x-axis, (b) parallel to the y-axis, and (c) at $35°$ to the x-axis.
4.3. Prove eqn (4.6) by setting the numerator in eqn (4.4) equal to zero and eliminating θ_2 by means of Snell's law. Calculate the Brewster angle for materials of refractive index 1.5, 1.6, and 1.9.

4.4. Prove that at Brewster incidence the reflected and refracted rays are mutually perpendicular.

4.5. Calcite has $n_o = 1.659, n_e = 1.487$. Calculate the thickness of a quarter-wave plate for wavelength 589 nm.

4.6. Show that the effect of a half-wave plate with its axis at an angle θ to the polarization direction of plane-polarized light is to rotate the plane of polarization through 2θ. (Resolve the incident field parallel and perpendicular to the plate axis.)

5. Image-forming instruments

Must get those old glasses of mine set right. Goerz lenses, six guineas.

5.1. Instrument design

In the design of an optical system for forming images, any or all of three factors may be important: (1) light-gathering power, or the capacity to form a bright image; (2) magnification; (3) resolving power, or the capacity to form sharp images of small detail. There may, of course, be other factors which matter in particular cases, such as weight, size, method of transporting an image to a particular place, stability to temperature changes, etc., but we shall not discuss these. The main factors (1), (2), and (3) are often interrelated in particular systems, and it is thus convenient to discuss them in relation to actual instruments.

5.2. Telescopes

Figure 5.1 shows the essentials of a refracting astronomical telescope. An objective lens, of diameter D and focal length f_1, forms a real image of stars and other astronomical objects at its focal plane. If the angle subtended

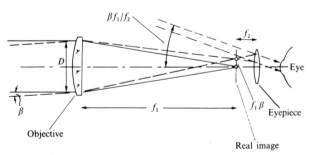

FIG. 5.1.   A refracting astronomical telescope. The objective and eyepiece are shown conventionally as single-component lenses. Rays from two stars are shown, one on axis (full lines) and the other an angle β away from the axis (broken lines).

between two of the stars is β then from Chapter 2 their images are separated by a distance βf_1. These real images are recorded by a photographic plate or other detecting system, or alternatively they may be viewed by the eyepiece. The eyepiece is a system of focal length f_2. It is used with the object, i.e. the real image of the star field, at its first focal plane. Thus it forms an image of the stars at infinity, so that they are viewed by the relaxed eye (probably equipped with glasses in the case of an elderly astronomer). The pair of stars now subtends an angle $\beta f_1/f_2$, so that the angular magnification is f_1/f_2. Thus for physical detectors the magnification is determined by a scale factor, according to which an angle β between objects at infinity corresponds to a distance βf_1 in the image plane, whereas for visual observation we use the angular magnification f_1/f_2.

The light-gathering power depends on conflicting factors. Image-recording detectors, in the sense used in Section 1.4, all have a certain minimum size of image which they can 'see'. Thus, suppose that by some means we produce an extremely small point image, say 250 nm across, on a photographic emulsion.† The image patch recorded by the emulsion will actually be much larger, owing to scattering and diffusion of light in the emulsion and other effects. Such an image produced by a negligibly small light patch is called a *point spread function*, and its size can be used to characterize the limiting performance of the detector. The concept is applicable to all detectors. In the normal human eye the point spread function projected back into the outside world corresponds to an angular subtense of about 0.0003 rad or a distance of 1 mm at 3 m; for a television camera it corresponds to the scan-line size; for photographic emulsions it can range from 500 nm (for special emulsions for holography or spectroscopy) to about 0.02 mm for very high-speed panchromatic emulsions.

Returning now to the telescope, if the star image is smaller than the point spread function of the emulsion, the light-gathering power must be proportional to D^2, since it is simply a question of the total of light flux collected. On the other hand, if the image is larger than the point spread function, as in the case of a planet or a nebula, the light-gathering power is measured by the flux per unit area falling on the emulsion, and then from Section 2.6 it is proportional to D^2/f^2. In considering light-gathering power for visual observation we have to ask whether the pupil of the eye admits all the light collected by the telescope. It can be seen by following rays through the system that the eyepiece forms an image of the aperture of the objective at a point to the right of the whole system where pencils from a star away from the axis cross the axis. In this context the aperture of the objective is the *entrance pupil*, and this image of it is the *exit pupil*. Clearly the eye must have *its* pupil roughly at the exit pupil of the telescope in order to see all the field of view. Thus the light-gathering power for visual purposes depends (1) on whether the exit

† This is about the smallest point image which can be produced in the optical region, as will be seen in Section 5.4.

pupil of the instrument is larger or smaller than that of the eye, and (2) on whether the star images are larger or smaller than the point spread function of the eye. This argument is taken further elsewhere (see e.g. Chapter 9 of Welford (1962)). The above discussion shows that the photometry of optical systems can be a complex topic.

A classical problem in astronomy is the resolution of double stars or similar close objects. On the basis of geometrical optics alone there need be no limit to resolving power; we merely have to make a telescope with adequate magnification and light-gathering power and with perfectly corrected aberrations. However, according to physical optics, there is a limit. We regard the two stars to be resolved as point sources of equal brightness and not, of course, coherent with each other, since they are in fact separate thermal light sources. The telescope aperture, i.e. the rim of the objective, limits the size of plane wavefronts accepted from one of the stars and thus diffraction occurs at the aperture. The objective itself then brings the far-field diffraction pattern to a convenient place for viewing—the focal plane. In other words, the image of a point object according to physical optics is the far-field diffraction pattern of the aperture of the objective. We have already described this image in Section 3.4 (eqn (3.16) and Fig. 3.13), and we know that it is a diffuse patch of light of which the angular size depends on the diameter of the diffracting aperture and on the wavelength of the light. From the present point of view we can call it the point spread function of the objective. Obviously the resolution of the optical system, leaving aside the effect of the detector, is determined by the size of this point spread function, since the image of the double star consists of two overlapping point spread functions. From the discussion in Section 3.4 the angular half-width of a point spread function is of order of magnitude λ/D, and this therefore is roughly the limiting angular separation at which the two stars can just be seen separate through the telescope, i.e. it is the *angular resolution limit*. It is traditional to take the resolution limit as the separation at which the central intensity maximum of one point spread function falls on the first dark ring of the other (as in Fig. 5.2), and from Problem 3.10 this gives for the angular resolution limit

$$\beta_{min} = 1.22\lambda/D. \tag{5.1}$$

In practice the difference between 1.22 and unity is rarely significant. Also in practice, the resolving power of ground-based astronomical telescopes with objective diameters greater than about 100 mm is limited by atmospheric turbulence which causes non-uniformity of refractive index in the atmosphere; then the originally plane wavefronts of uniform amplitude from a star are distorted and this causes the star image to wander, broaden, and vary in intensity with time. Because of this effect the angular resolution of such telescopes cannot be better than about 1 arcsecond when they are used for direct image formation and the very large diameter objectives, up to 5 m, are for light-gathering power, not for resolution.

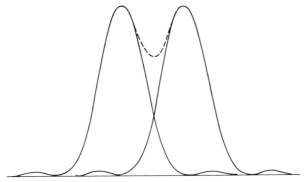

FIG. 5.2.   Two star images of equal intensity which are conventionally a just-resolvable distance apart. The angular separation of the stars is 1.22 λ/D where D is the diameter of the telescope objective. The sum of the two images is indicated in broken line.

The above discussion shows how the factors listed in Section 5.1 can enter into any given problem. To summarize: (1) we must have enough light-gathering power to record the required event in a suitable time, e.g. up to several hours in astronomy, less than a nanosecond in the study of pulsed lasers, or anything in between; (2) there must be adequate magnification to ensure that the detector can separate all the detail in the image which is (3) resolved by the optical system. We must always make a clear distinction between (2) and (3). For a telescope the resolution depends only on the diameter of the objective and it is independent of the focal length. On the other hand, again in the case of the telescope, the scale of the star picture on the photographic plate depends on the focal length but *not* on the diameter of the objective. A similar distinction can be drawn for most optical instruments.

Light-gathering power is the most important factor in modern astronomy. For technical reasons concerned with chromatic aberration and with the manufacture of optical glass all large telescopes have mirror objectives. Figures 5.3–5.5 show three widely used types. The Newtonian telescope (Fig. 5.3) is simply a large concave paraboloid of revolution. It is easily shown

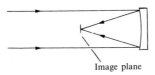

Image plane

FIG. 5.3.   The Newtonian telescope. The photographic plate or other detector is placed at the image plane. (Newton's instrument was, of course, used visually with a small plane mirror to permit viewing from the side of the telescope tube: modern telescopes with a paraboloidal primary mirror are always called Newtonian.)

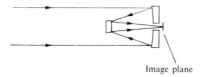

Image plane

FIG. 5.4.   The Cassegrain telescope. The convex mirror, called the secondary, has a hyperboloidal shape if the primary is a paraboloid. The system has a relatively long focal ratio ($F/8$ to $F/11$) to match a spectroscopic system attached to the telescope.

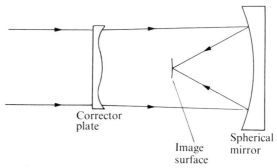

Corrector plate

Image surface

Spherical mirror

FIG. 5.5.   The Schmidt camera. The essential feature is the corrector plate with a non-spherical surface. The main mirror is spherical, and the corrector plate provides the aberration correction which the main mirror would have if it were a paraboloid. However, since the corrector is at the centre of curvature of the mirror the same correction applies at all field angles, and thus the camera has a large, well-corrected field of view.

that this brings incident rays parallel to the axis exactly to a point focus at the (geometrical) focus of the generating parabola; the detector, e.g. a photographic plate or a photoelectric image-intensifier, is placed at the image plane. The Newtonian is the system with greatest speed for direct image recording, since the mirror can have a focal ratio of about $F/3$.† Sometimes a spectrographic system has to be added to the telescope, and then it is better to have a longer or larger focal ratio, say $F/8$ to $F/11$, in order to match that of the spectrograph. The Cassegrain system (Fig. 5.4) is then used. It is not usually a separate telescope, but is obtained by adding an auxiliary convex mirror to a Newtonian primary mirror; the larger focal ratio is thus obtained without an increase in overall length of the system.

Both the Newtonian and Cassegrain telescopes have very small angular fields of view, owing to off-axis aberrations. On a large telescope the well-corrected field may be only about a minute of arc. The Schmidt telescope

† An objective is said to have a focal ratio or F-ratio of F/N if its focal length is N times the diameter of its aperture or entrance pupil. Thus a small focal ratio is photographically very fast.

(Fig. 5.5) can have an angular field of a few degrees at about $F/2.5$ for apertures exceeding 1 m; it is normally used photographically (and therefore usually called the Schmidt *camera*) for rapid, large-scale surveys of the sky.

5.3. The human eye

We have to describe the eye both as an optical system and as a detector, since it is used together with other optical systems. Figure 5.6 is a very simplified diagram of the eye. Most of the refracting power is in the front surface of the

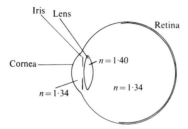

FIG. 5.6.   The human eye. This is a very schematic and simplified diagram. Most of the lens effect is due to the sharply curved front surface of the eye, the cornea. The iris is the aperture stop of the eye.

cornea, and the main function of the lens is to vary the focus so as to be able to see clearly over a range of distances. Thus the eye is like a camera, forming an image of a distant scene on a light-sensitive surface, the retina. The retina is an array of light-sensitive cells which communicate with the brain *via* a complex network of interconnected nerve cells. The normal eye can adjust its focus or *accommodate* to form sharp images of objects at distances from infinity to about 250 mm.† The accommodation is not done as in a camera, by changing the distance between the lens and the detector, but by varying the lens curvatures by muscular control. The iris varies in diameter between about 2 mm and 8 mm, depending on the average brightness of the scene being viewed. The sensitivity of the visual channel from the retina to the brain also varies (*adaptation*), so that the eye can be used over a very wide range of light intensities, more than 8 orders of magnitude for diffusely illuminated scenes. A very faint flash consisting of only a few photons of green light striking a single receptor in the retina can be detected under suitable conditions, and at the other end of the scale a flux of 0·1 mW on a single receptor can be tolerated for about 0.1 s, so that on this basis the range of sensitivity of the eye is about 14 orders of magnitude. This range is obtained, as in most forms of

† But the range of accommodation decreases in old age. Many people have a different range, and they have to wear glasses in order to add or subtract power to shift the range to the 'normal'. Children can often accommodate to much closer than 250 mm.

sensory perception, by an approximately logarithmic response, i.e. a given *difference* in sensation corresponds to the same *ratio* between light signals at any part of the range. In many physical measurements a linear relation between the quantity measured and the indication is desired, but the great compression of range produced by a logarithmic response is often very useful.

As noted in Section 5.2, the angular resolution of the eye is about 1 min of arc (0.0003 rad), depending on the conditions. This figure is used in determining the required magnification of an optical instrument. Thus if a telescope can *resolve* detail of 1 sec of arc we have to make the eyepiece magnify this detail enough to subtend 1 min to the eye (in practice 2 or 3 times more). The magnification is calculated as in Section 5.2.

We must stress that the above description of the eye is very incomplete and lacking in detail. The responses of the eye to varying light levels, to fine detail, and to different wavelengths are very complicated, and are by no means fully understood. Our description is intended merely as a sketch to suggest how the eye is coupled to other optical systems.

5.4. The microscope

A microscope is essentially an elaborate magnifying glass. If we use a lens of focal length f to form an image at infinity of an object of size η (as in Fig. 5.7), the magnified object appears to subtend an angle η/f. On the other hand, if we

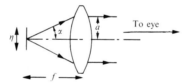

FIG. 5.7.   A magnifying glass. The object is (approximately) at the front focus, so that the image is seen by the relaxed eye at infinity and it appears to subtend an angle η/f. In practice the magnification is not very dependent on the exact position of the object.

use the unaided eye the object must be at least 250 mm away because we cannot accommodate nearer than this distance (Section 5.3), so the object then appears to subtend an angle $\eta/250$. Thus the magnification is $250/f$, provided f is in millimetres. We must also ensure that we can resolve the magnified detail. Thus from Section 5.2 the point spread function of the lens appears to the eye to subtend an angle $\lambda/2a$, where $2a$ is the diameter of the magnifier, or, turning this into a distance in the object space, the point spread function is $\lambda f/2a$ across, and this must be smaller than the detail to be resolved. Also a must be small enough to ensure that all the rays get into the pupil of the eye. The quantity a/f is roughly the semi-angle of the cone of rays collected by the lens from the axial object point, and it must be made as large as possible to resolve small detail, since the size of the point spread function is proportional to f/a. In fact it is found (Welford 1962) that it is the sine of this angle which matters, and the accepted form of the resolution limit is

$$\eta_{min} = 0.5\lambda/\sin \alpha. \qquad (5.2)$$

The precise value of the numerical factor depends in a complicated way on the conditions of illumination of the object and on just how we define 'resolution', but the value 0.5 is adequate for most practical purposes.

Since the diameter $2a$ must be less than, say, 4 mm in order to match the pupil of the eye, it can be seen that we have to use very short focal length lenses in order to get high resolving power, and focal lengths of 2 mm or less are used with angles α up to about 60°. The lenses have to have many components to keep the aberrations small, and it is then found to be impossible to get the eye close enough to the lens to see a reasonable field of view. Thus the so-called compound microscope was developed, as in Fig. 5.8. An objective with much

Object plane

Primary
image plane

FIG. 5.8.   Principle of the compound microscope. The objective has a very small focal length so that the primary image is highly magnified, and it has a large collecting angle α for high resolution. The primary image may be recorded directly on a physical detector, e.g. a photographic emulsion, or it may be viewed through an eyepiece.

the same characteristics as we postulated for the magnifier in Fig. 5.7 is used to form an enlarged real image of the object. This real image can be recorded photographically or by means of any other physical detector, just as for the primary image in an astronomical telescope; or alternatively we can form a virtual image of it at infinity with an eyepiece and simply look at that, as in conventional microscopy. Whatever mode of detection is used we again note that the function of resolution depends on the objective and its collecting angle, but not on the details of the optical system which follows. Since the wavelength appears in eqn (5.2) in the numerator we can also gain resolution by going to shorter wavelengths. Little has been done with ultraviolet-light microscopy, but great gains have been made in electron microscopy. The de Broglie wavelength of electrons of 1 MeV energy is about 10^{-12} m (*Radiation and quantum physics*, OPS 3) and the collecting angle of electron lenses used as microscope objectives is of the order of 10^{-3}, so that the latest electron microscopes are approaching the resolution of intermolecular distances. However, the collecting angle is limited to that small value by aberrations which seem to be in principle irreducible, whereas optical microscope objectives can be made practically aberration-free for values of α up to $60°$.

In eqn (5.2) λ in the numerator is the wavelength in the medium containing the object. Thus if the object is embedded in a medium of refractive index n then $\lambda = \lambda_0/n$, where λ_0 is the vacuum wavelength of the light. Equation (5.2) then takes the form

$$\text{resolution limit } \eta_{min} = 0.5\lambda_0/n \sin \alpha. \qquad (5.2a)$$

and we see that a gain in resolution is obtained by so embedding or immersing the object in a medium of high refractive index. This is the principle behind the oil-immersion microscope objective. The quantity $n \sin \alpha$ is called the *numerical aperture* (NA), and it is quoted on microscope objectives as a measure of resolving power.

5.5 Images of extended objects

Telescopes and microscopes have relatively small fields of view and the aberrations can be made small enough over these fields to justify taking the point spread function as the ideal Airy pattern (Fig. 3.13). This is not so with wide field systems such as objectives for film or television cameras, and then the point spread function formed with aberrations can be very different from the Airy pattern. Thus a 50 mm $F/1.8$ camera lens has the large aperture $F/1.8$ so that short exposures can be used. If the lens had no aberrations its point spread function would be that corresponding to a convergence angle of about $20°$, i.e. it would be less than 1 μm in half-width, and this would be unnecessarily small for ordinary photographic emulsions. If an average is taken over a reasonable wavelength range, so that fine diffraction structure is smoothed out, the point spread function might appear as in Fig. 5.9. Aberrations will often cause the point spread function to be unsymmetrical as suggested in the figure.

We can obtain an expression for the point spread function from eqn (3.11b), since in the present context this equation gives the complex amplitude in the point spread function. Equation (3.11b) refers to a diffraction pattern in the far-field, and so in order to bring it to the focal plane of the camera lens we replace the angular coordinates u and v by ξ/f and η/f, where ξ and η are linear coordinates in the focal plane and f is the focal length. In Section 3.4 we introduced the pupil function $F(x, y)$, which defines the pupil area and allows us to use the Fourier transform formalism. The pupil function can be used to define a pupil of any irregular shape, e.g. the pentagonal iris diaphragm used in some cameras with automatic exposure control, and in advanced work it is also made to include the effect of

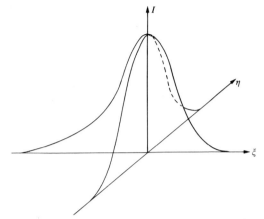

FIG. 5.9.   A point spread function. The height of the solid at (ξ, η) is proportional to the light intensity in the point spread function.

aberrations. Then if $I(\xi, \eta)$ is the light-intensity distribution in the point spread function we have, from eqn (3.11b)

$$I(\xi, \eta) = \left| \frac{1}{A} \int\limits_{-\infty}^{\infty}\!\!\int F(x, y) \exp \left\{ \frac{i2\pi}{\lambda f} (\xi x + \eta y) \right\} dx \, dy \right|^2. \qquad (5.3)$$

We cannot evaluate this expression until we know the form of the pupil function $F(x, y)$, and often the integration has to be done numerically, but we suppose this to be done.

We next consider how the image of an extended object such as a bright disc or square is built up from individual point spread functions. We assume the object to be incoherently illuminated, i.e. the light from any one point of the object cannot interfere with that from any other point, so that we obtain the effect of overlapping point spread functions by adding their *intensities*. This will be so if the object is self-luminous, e.g. a hot filament or a gas discharge, or if it is illuminated by a non-monochromatic source of large enough size. Suppose then that the distribution of light intensity in the object is given by $O(\xi, \eta)$. We use the same coordinates (ξ, η) as in the image space to denote points which are object and image in Gaussian approximation, by simply rescaling the object coordinates according to the magnification of the optical system. An element of the object, say $O(\xi', \eta') \, d\zeta' \, d\eta'$, produces a point spread function which redistributes the light from the element over the image plane. Thus a point $P(\xi, \eta)$ receives light from all the spread functions, and in particular from the spread function at (ξ', η') it receives the contribution

$$I(\xi - \xi', \eta - \eta')O(\xi', \eta') \, d\xi' \, d\eta',$$

as in Fig. 5.10. The total effect at P is obtained by summing over all the points

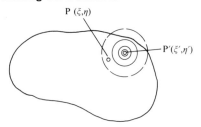

FIG. 5.10.   The image of an extended object. The point spread function centred at $P'(\xi', \eta')$ contributes some light intensity at $P (\xi, \eta)$.

(ξ', η') of the object; then provided the spread function has the same form at all these points we have

$$O'(\xi, \eta) = \iint I(\xi - \xi', \eta - \eta')O(\xi', \eta')\, d\zeta'\, d\eta', \tag{5.4}$$

and this is the light-intensity distribution in the image $O'(\xi, \eta)$ of the object $O(\xi, \eta)$. Thus we see that the image is obtained as the convolution of the object and the point spread function (see Appendix). Convolution is the mathematical representation of a physical process in which a sharply defined input is spread to produce a blurred output. For example, in a communication channel the output signal is the convolution of the input signal and the *impulse response*. This latter is the response of the channel to a delta-function input signal, and so it corresponds to the point spread function in an optical system.

We can formally write the object and image intensity distributions $O(\xi, \eta)$ and $O'(\xi, \eta)$ and the point spread function $I(\xi, \eta)$ as the inverse Fourier transforms of certain functions $o(s, t)$, $o'(s, t)$, and $L(s, t)$,

$$O(\xi, \eta) = \iint o(s, t) \exp \{i2\pi(s\xi + t\eta)\}\, ds\, dt, \tag{5.5}$$

$$O'(\xi, \eta) = \iint o'(s, t) \exp \{i2\pi(s\xi + t\eta)\}\, ds\, dt, \tag{5.6}$$

$$I(\xi, \eta) = \iint L(s, t) \exp \{i2\pi(s\xi + t\eta)\}\, ds\, dt, \tag{5.7}$$

and the convolution theorem (Appendix) tells us that

$$o'(s, t) = o(s, t)L(s, t). \tag{5.8}$$

The physical significance of eqns (5.5)–(5.7) is that the object and image are expressed as sums (i.e. integrals) of sinusoidal components, s and t are *spatial frequencies*, e.g. numbers of lines per millimetre, and eqn (5.5) means that the periodic component in $O(\xi, \eta)$ with spatial frequency components s and t in the x- and y-directions has the amplitude $o(s, t)$ ds dt. Then according to eqn (5.8) the amplitude of the periodic component in the image having the same pair of spatial frequencies is obtained by multiplying the amplitude of the object component by the factor $L(s, t)$, the Fourier transform of the point spread function (eqn (5.7)). This function $L(s, t)$ is called the *optical transfer function*, usually abbreviated to OTF, and its role in an optical system is analogous to that of the transfer function of an electrical channel such as an amplifier. The form of the OTF, i.e. its numerical values, depends on the form of the point spread function, but it can be shown that whatever the form of the OTF a single-frequency object or sinusoidal grating forms a similar (i.e. also sinusoidal) image, but with different contrast. The reduction in contrast as a function of spatial frequency is used as a measure of quality for optical systems which are to be used for imaging extended objects in incoherent illumination. Figure 5.11 shows how images of sinusoidal objects of different

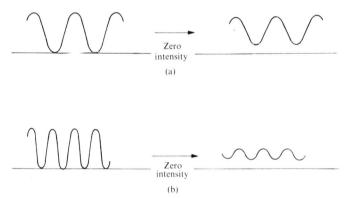

FIG. 5.11. Contrast transfer for sinusoidal objects. (a) and (b) objects of low and high spatial frequencies respectively, showing greater reduction in contrast of the image, i.e. smaller relative swing in intensity, for (b).

spatial frequencies are formed. The contrast is usually less for higher spatial frequencies and there is no image contrast at all, i.e. no modulation, for spatial frequencies above a certain limit corresponding roughly to a grating with lines spaced at the resolution limit (eqn (5.2)).

Problems

5.1. What is the theoretical angular resolution of telescopes with objective diameters of 100 mm, 1 m, and 5 m?

5.2. A telescope with a 50 mm aperture and 500 mm focal length objective is used for stellar photography. Estimate the size of a star image on the plate and calculate the plate scale in radians per millimetre.

5.3. The telescope in Problem 5.2 is to be used visually. What focal length would be required for the eyepiece in order to take advantage of the theoretical resolution of the objective?

5.4. Plot a graph of the light intensity across the centre line of two star images of equal intensities when the centres are separated by a distance equal to the radius of the first dark ring of the Airy pattern.

5.5. What is the resolution limit of a microscope objective of numerical aperture 0.65, and what overall magnification of the microscope is needed to take full advantage of this resolution?

5.6. Suggest a microscope NA and a magnification suitable for visual study of (a) red blood cells 7 μm in diameter and (b) grains of silver halide 1 μm in size in a photographic emulsion.

6. Interferometers and spectroscopes

Red rays are longest. Roygbiv Vance taught us: red, orange, yellow, green, blue, indigo, violet.

James Joyce: Ulysses

6.1. Young's experiment; spatial coherence

Figure 6.1 shows a simple way to produce interference effects. There are two pinholes in a screen placed in a collimated beam; the light from each pinhole spreads by diffraction into a cone of semiangle approximately λ/d, where d is the pinhole diameter, (see Section 3.4). Thus the pinholes are secondary sources producing diverging spherical wavefronts; these interfere where they overlap, since they have come from the same original source, and they produce interference fringes as in the figure. A simplified version of this experiment was carried out by Thomas Young in 1804. The experiment led to the general acceptance of the theory that light is a wave phenomenon.

Young's experiment can be used to illustrate the concept of coherence between light beams. If the source in Fig. 6.1 is a helium–neon laser it is not essential to have a pinhole at the focus of the collimator and the fringes will have good contrast or visibility if the two secondary pinholes are equal in diameter. If we use a thermal source such as a sodium lamp we find experimentally that the collimator pinhole must be restricted in size for fringes of good contrast to be formed. We can see why this is by an argument similar to that in Section 3.1. Let the source pinhole have diameter d, let the

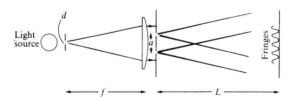

FIG. 6.1. A version of Thomas Young's interference experiment. The collimator is not essential. It is put in to make the phase relationships between the two pinholes clearer.

collimator focal length be f, and let the two secondary pinholes be a distance a apart. The spacing of the fringes formed by light from a given point in the source is $\lambda L/a$, where λ is the wavelength and L is the distance from the pinholes at which the fringes are observed. The phase difference at the screen between disturbances from source points on either side of the pinhole is $(2\pi/\lambda)ad/f$, so that the different source points form fringe systems displaced laterally by the fraction of a fringe $ad/\lambda f$, as in Fig. 6.2. Thus if $ad/\lambda f$ is of order

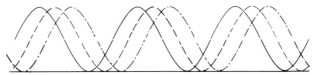

FIG. 6.2.   Superposed fringes produced in Young's experiment by a source pinhole of finite size.

of magnitude unity or greater there will be more or less uniform illumination, i.e. no fringes will be seen. In other words, we must have

$$ad/f \leqslant \lambda/4 \qquad (6.1)$$

for fringes of good contrast to be formed.

This equation has more than one interpretation. It tells us how small the source pinhole must be in order to get good fringe contrast: it must be smaller than $\lambda f/4a$, and it is then said to be a '*diffraction-limited pinhole*'. Equation (6.1) also tells us the distance between points on the screen within which the illumination is substantially coherent. This distance is about $\lambda f/4d$, i.e. this is the order of size of a '*coherence patch*' from the source pinhole subtending the angle d/f. Thus the experiment illustrates the principle stated in Chapters 1 and 3 that there is coherence between light illuminating two different points if interference fringes of good contrast can be formed between beams of light from the two points. This is made quantitative in more advanced treatments by defining a degree of *partial coherence* according to the actual contrast of the fringes (Born and Wolf 1965). We see from eqn (6.1) that as the source gets bigger the coherence patch gets smaller.

We can use eqn (6.1) in yet another way. If we measure the size of the coherence patch from an inaccessible source we have a measure of the size of the source. This is the principle of the stellar interferometer invented by A. A. Michelson. Light from a star is collected by two mirrors of variable separation, a, as in Fig. 6.3, and the two beams are brought together and made to interfere. Let the angular subtense of the diameter of the star be β; then from eqn (6.1) the fringes will have good contrast provided a is less than $\lambda/4\beta$, but as a increases beyond this value the contrast will decrease. The detailed theory shows that the fringe contrast falls to zero, i.e. the fringes disappear,

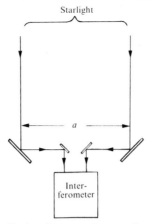

FIG. 6.3.   Michelson's stellar interferometer. Beams a distance a apart on the ground but coming from the same star are made to interfere. The details of the interferometer do not matter in principle. The contrast of the interference fringes depends on the angular diameter of the star.

when a is approximately λ/β, or, more precisely for a star of circular shape and uniform brightness, $1.22\lambda/\beta$. In Michelson's instrument the mirror separation a could take values up to about 6 m. The details of the arrangement by which the interference fringes are produced are in principle irrelevant to the measurement: we merely have to make the beams interfere. The same principle is applied in measuring the diameter of radio stars, but the 'interference' is arranged by mixing the radio signals collected by antennae at suitably variable spacings. The wavelengths are in the centimetre range, and the separations are of the order of a kilometre.

6.2. Michelson's interferometer; temporal coherence

In Section 3.1 we discussed briefly interference between beams reflected at the two surfaces of an oil film on water or a similar thin layer. The detailed theory of these effects is complicated by multiple reflections to and fro in the film. The principles can be seen more clearly in an apparently more complicated apparatus, *Michelson's interferometer* (not to be confused with his stellar interferometer), shown in outline in Fig. 6.4. In principle, this is merely a device for studying interference between coherent beams reflected from two parallel or nearly parallel surfaces, but in order to avoid multiple reflections the surfaces are not placed almost in contact, as in the oil film. The surfaces are the two plane mirrors M_1 and M_2. The beam-splitter, or semi-reflecting and semi-transmitting mirror, is adjusted to make the image of M_2 appear

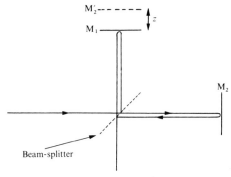

FIG. 6.4.   Michelson's interferometer (not to be confused with the stellar interferometer of Fig. 6.3), M_2' is the apparent position of M_2 as seen in the beam-splitter.

parallel to M_1 and at a distance z from it, as seen by a detector (e.g. the eye) looking into the system as shown. We consider interference between collimated beams from a source at infinity having a finite angular subtense; this could be arranged as in Fig. 6.5, where the source, e.g. a mercury lamp, is at the focus of a collimator of focal length f_1 and has a diameter $2a$. The interference effects are to be observed at the far-field, since this is where the individual source points are imaged, and for this we use an objective of focal length f_2 and observe at its focal plane.

A point of the source at a distance ρ from the axis of the collimator produces a collimated beam inclined at an angle ρ/f_1 to the axis of the

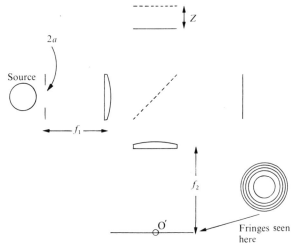

FIG. 6.5.   Formation of circular fringes in Michelson's interferometer.

interferometer. This beam therefore meets an effective plate of air of thickness z at an angle of incidence ρ/f_1, and from eqn (3.4) the optical path difference between the two beams reflected back from M_1 and the image of M_2 is

$$W = 2z \cos (\rho/f_1). \qquad (6.2)$$

At the centre of the far-field pattern this path difference has the value $2z$, and if this is an integral number of wavelengths the centre will be bright.† Going out from the centre, W decreases because of the cosine factor in eqn (6.2), and each time it decreases by one wavelength we reach another bright fringe. The fringes must be circular, since W depends only on the angle of incidence ρ/f_1, not on the azimuthal angle. It is easy to show that the radius of the Nth fringe is proportional to $N^{1/2}$, as for Newton's rings.

Consider only the centre of the fringe system $0'$, and suppose the distance z to be varied steadily. Then, as the path difference $W = 2z$ varies, $0'$ will be alternately bright and dark, and the intensity there will be proportional to

$$1 + \cos (4\pi z v/c), \qquad (6.3)$$

where v is the frequency of the light. A small enough detector at $0'$ would record this as a fringe pattern or *fringe function* in the variable z.

So far we have assumed the light to be monochromatic with frequency v. If this is not so (e.g. we might be using a source with a broad spectrum line, such as a high-pressure mercury lamp, or perhaps a continuous source, such as a filament lamp) we suppose that the proportion of power in the light beam between frequencies v and $v + dv$ is $G(v)\, dv$. Thus $G(v)$ is proportional to the light intensity seen through a prism or other spectroscopic system. The fringe function for the frequency band dv is then

$$G(v)\{1 + \cos (4\pi z v/c)\}\, dv$$

and the total fringe function for the light of all frequencies added together is obtained by integrating with respect to v,

$$g(z) = \int G(v) \{1 + \cos (4\pi z v/c)\}\, dv. \qquad (6.4)$$

The physical meaning of this equation is that we are adding together individual fringe functions of different fringe spacings. The fringe spacing in the z domain for frequency v is $c/2v$. Thus these fringe systems start in phase at zero path difference ($z = 0$) and they gradually get out of phase as z increases, so that the contrast of the fringes falls. For nearly monochromatic light, i.e. a small range of frequencies, the contrast is good over a large path difference, and for 'white' light only a few fringes can be detected with measurable contrast. Thus the interferometer can be used to estimate the narrowness of a spectrum line. Michelson himself used it in this way in 1892 to show that the

† Here, as elsewhere, we ignore complications due to phase-change effects on reflection (see Section 3.1).

red cadmium line of wavelength 643.8 nm is very narrow and is therefore suitable for standardizing the metre in terms of wavelengths.

Returning to eqn (6.4) we see that the right-hand side is the sum of a constant (the integral over the spectrum of $G(v)$) and the cosine Fourier transform of $G(v)$. Thus if the fringe function $g(z)$ is recorded we can obtain the spectrum of the light by calculating the Fourier transform. This is the principle of *Fourier-transform spectroscopy*, which has become a standard technique in many fields within the last two decades. Michelson actually determined several spectrum line profiles in this way over 80 years ago.

The fall in contrast of the fringe function with increase of path difference $2z$ can be regarded also as decreasing coherence between the beams returning from the mirrors. Thus the path difference at which the contrast falls to some chosen value is a measure of the coherence length (Section 1.5) and the corresponding coherence time is $2z/c$. These are therefore measures of the length of wave-train which is reasonably correlated with itself or which is approximately sinusoidal with the same frequency and amplitude along its length. Thus coherence length and spectral composition are two different aspects of the same physical phenomenon.

The Michelson interferometer of Figs. 6.4 and 6.5 has many other applications (see e.g. Born and Wolf 1965), but that which we have described has the greatest importance in basic physics.

6.3. Prisms and gratings as dispersing elements

The conventional spectroscope disperses the different wavelengths into different angular directions. The dispersing prism is still used essentially as in Isaac Newton's experiment to produce a spectrum. Figure 6.6 shows the principle of a *prism spectrograph*, used for photographing a spectrum. The deviation of a beam by a prism depends on the wavelength, since the refractive index of all materials varies with wavelength: since the deviation also depends on the angle of incidence of the light (as can be shown by a rather laborious application of Snell's law) it is necessary to collimate the light. The spectrum is formed at infinity with respect to the prism, i.e. in the far-field, and therefore it is brought to a focus by the camera objective. Thus the complete system

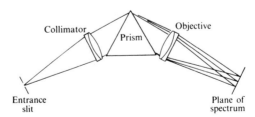

FIG. 6.6.   Principle of the prism spectrograph.

consists of a slit at the focus of the collimator to define the direction of the incident beam, the dispersing prism, the camera objective to focus the dispersed beams, and the photographic plate or other image-recording detector. Figure 6.6 shows only the principle of the simplest kind of spectrograph; there are many variations for different purposes.

The mode of action of a *diffraction grating* is not quite so obvious as that of a prism. Figure 6.7 shows a grating consisting of narrow slits in an opaque

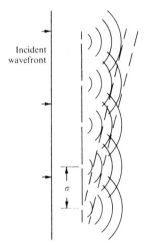

Incident
wavefront

σ

FIG. 6.7.   First-order diffracted wavefronts as the envelopes of secondary waves from the slits or 'rulings' of a diffraction grating.

screen with spacing σ between the slit centres. If a collimated beam of light is incident normally from the left each slit diffracts light as in Chapter 3, and, if the slits are narrow enough, the diffracted light spreads out over a range of angles, as indicated. The diffracted beams from each slit interefere, since they originated from the same collimated beam. In the far-field all diffracted beams will be in phase in the direction of the original incident beam, but there can be other directions in the far-field in which beams from neighbouring slits are one wavelength out of phase with each other. Such a diffracted beam is indicated in the figure.

Figure 6.8 shows how we can calculate the directions of this and other diffracted beams: the ray and wavefronts are indicated in inverted commas because they do not exist in the near field, but the construction gives the direction of the diffracted ray and wavefronts. For the direction α' we must have

$$\sin \alpha' = \lambda/\sigma. \qquad (6.5)$$

There may also be other diffracted beams with two, three, or more

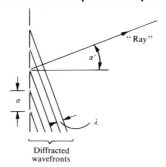

FIG. 6.8.   Calculation of the angle of diffraction. The 'ray' is the common normal to the 'wavefronts', and these are the envelopes of the actual diffracted wavefronts, as in Fig. 6.7.

wavelengths path difference between the waves from successive slits, and for these we should have

$$\sin \alpha' = M\lambda/\sigma \; (M \text{ an integer}). \tag{6.6}$$

Since in eqns (6.5) and (6.6) the angle of diffraction depends on the wavelength, a collimated beam of white light incident on the grating will be spread into a spectrum. In fact there will be several spectra corresponding to the different orders M, and there will be an undispersed zero order, the light which travels on undeviated. This is indicated in Fig. 6.9.

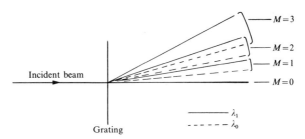

FIG. 6.9.   Spectra of different orders formed by a grating: λ_1 is greater than λ_0.

Actual spectroscopic gratings are not made as in Fig. 6.7. The distribution of light flux between the spectra of different orders depends on the form of the individual slits or rulings; any periodic structure acts as a diffraction grating (e.g. a fabric of regular weave or the surface of a gramophone record), and in a spectroscopic grating the form of the ruling, i.e. the variation of transmission and phase across an individual slit or line, is designed to throw as much light flux as possible into a single diffracted order. Most gratings work in reflection

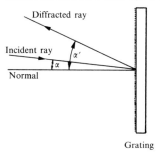

Grating

FIG. 6.10.   A plane reflection grating.

rather than in transmission, as in Fig. 6.10. By an extension of the argument used above we can show that for angles of incidence and diffraction α and α' the relation corresponding to eqn (6.6) is

$$\sin \alpha + \sin \alpha' = M\lambda/\sigma. \tag{6.7}$$

This equation gives the direction α' in which the Mth-order diffracted beam of wavelength λ goes for an angle of incidence λ. Obviously it reduces to the ordinary law of reflection for the zero-order beam.

Many grating spectroscopes use reflecting collimators and objectives rather than lenses because of the better aberration correction (*no* chromatic aberration) and greater wavelength range which can be used with mirrors. Figure 6.11 shows one simple design of *monochromator*, i.e. a system with a

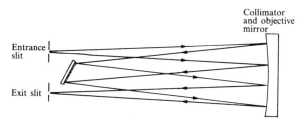

FIG. 6.11.   A grating monochromator. The grating is rotated about an axis normal to the plane of the diagram to scan the spectrum across the exit slit.

second slit at the plane of the spectrum. Different wavelengths are scanned across the exit slit when the grating is rotated. If the grating is formed on a concave surface the image-forming properties of the concave mirror are combined with the dispersion of the grating, and the spectrum is formed and focused by a single element, the *concave grating*, as in Fig. 6.12. If a complete spectrum is recorded photographically, as in the figure, the instrument becomes a *grating spectrograph*.

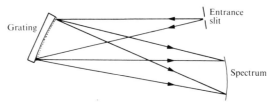

FIG. 6.12.   The concave diffraction grating arranged as a spectrograph.

6.4. Dispersion, resolution, and light-gathering power of prisms and gratings

The light-gathering power of spectroscopic systems depends on the area of the entrance slit and on the angular subtense of the collimator aperture in much the same way as for ordinary image-forming systems, but it also depends on the dispersion, i.e. the angular separation in the far-field per unit wavelength or frequency interval. The light-gathering power also depends on the nature of the detector, i.e. whether this is an image-recording system, such as a photographic emulsion in a spectrograph, or a total flux collector, such as a photomultiplier at the exit slit of a monochromator. In the simplest case, a monochromator with entrance and exit slits of equal width and with a dispersing element of area A, the flux transmitted in the wavelength interval $\delta\lambda$ is proportional to $\beta\delta\alpha A\delta\lambda$, where $\delta\alpha$ is the angular subtense of either slit along the direction of dispersion and β is the angular subtense of the height of either slit. This can be written

$$\beta\delta\alpha\cdot A\,\frac{\mathrm{d}\lambda}{\mathrm{d}\alpha'}\cdot\delta\alpha', \tag{6.8}$$

where $\mathrm{d}\lambda/\mathrm{d}\alpha'$ refers to the dispersing element. Equation (6.8) is a general result for light-gathering power applicable to any spectroscopic system with angular dispersion, i.e. where the different wavelengths in the light to be analysed are sent in different directions. We next have to obtain particular formulae for the angular dispersion $\mathrm{d}\lambda/\mathrm{d}\alpha'$ appropriate to the prism or grating to be used.

We calculate the angular dispersion of a grating by differentiating eqn (6.7) with respect to α',

$$\frac{\mathrm{d}\lambda}{\mathrm{d}\alpha'} = \frac{\sigma}{M}\cos\alpha'. \tag{6.9}$$

Thus for small angles of diffraction the dispersion of a grating is almost linear, since $\cos\alpha' \sim 1$ for small α', i.e. the wavelength found in the spectrum is directly proportional to the angle of diffraction.

Figure 6.13 shows the notation used for deriving the formula for the

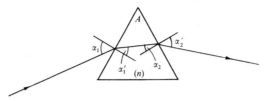

FIG. 6.13.   Notation for calculating the dispersion of a prism.

dispersion of a prism.   By differentiating Snell's law for both faces and eliminating the internal angles of incidence it can be shown that

$$\frac{d\alpha'_2}{dn} = \frac{\sin A}{\cos \alpha' \cos \alpha'_2},$$

or

$$\frac{d\alpha'_2}{d\lambda} = \frac{\sin A}{\cos \alpha'_1 \cos \alpha'_2} \cdot \frac{dn}{d\lambda}, \tag{6.10}$$

where $dn/d\lambda$ is the dispersion of the prism material. It can also be shown that in the symmetrical position, with $\alpha_1 = \alpha'_2$, the total angular deviation of the beam is a minimum, and for this *minimum deviation* position we have

$$\frac{d\alpha'_2}{d\lambda} = \frac{2 \sin \frac{1}{2} A}{\cos \alpha'_2} \cdot \frac{dn}{d\lambda}. \tag{6.11}$$

For both the diffraction grating and the prism the final image of a point in the entrance slit is formed as the far-field diffraction pattern of an aperture which may be either the rectangular outline of the prism or grating itself or the aperture of a lens or mirror used to bring the far-field pattern to a focus. Thus an indefinitely narrow entrance slit illuminated with perfectly monochromatic light would still produce a spectrum line of a certain finite angular subtense. This, by the same reasoning as we used in Section 5.2, must be of order of magnitude λ/D, where D is the width of the prism or grating aperture across the direction of dispersion. The criterion for the just-resolvable separation of two wavelengths λ and $\lambda + \delta\lambda$ therefore that their directions shall be separated by this angle, as in Fig. 6.14. The *resolving power* of a dispersing element is conventionally defined as the number

$$\lambda/\delta\lambda, \tag{6.12}$$

and we can find this for a grating by putting $d\alpha' = \lambda/D$ in eqn (6.9). We obtain

$$\frac{\lambda}{\delta\lambda} = \frac{MD}{\sigma \cos \alpha'},$$

but, recalling that D is measured across the width of the beam diffracted to the

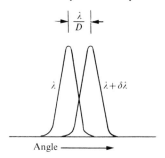

FIG. 6.14.   The concept of spectroscopic resolving power.

far-field as in Fig. 6.15, we see that $D/\sigma \cos \alpha'$ is the number N of rulings on the grating. Thus we have

$$\lambda/\delta\lambda = MN, \qquad (6.13)$$

or, *the resolving power of a diffraction grating is the product of the number of rulings and the order of diffraction.*

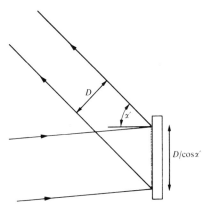

FIG. 6.15.   Resolving power of a plane grating.

Modern spectroscopic gratings for the visible region of the spectrum can be 50–300 mm wide and can have 300–1000 rulings per millimetre (but these are not absolute limits), so that a resolving power of order 3×10^5 is theoretically possible in the first-order spectrum. For comparison a spectrum line from a low-pressure gas discharge might have a width due to Doppler broadening of about 10^{-6} of its wavelength, so that its structure could be nearly resolved by such a grating.

We can obtain the resolving power of a prism by similar reasoning to that used above, but it is easier to proceed from first principles. Figure 6.16 shows a collimated beam of wavelength λ traversing a prism, with a plane wavefront Σ of the transmitted beam. If the wavelength is decreased by $\delta\lambda$ the refractive index will increase by a corresponding amount δn on account of the dispersion of the glass; the optical path length through the base of the prism will increase by $\delta n \cdot d$, where d is the length of the path through the base, and

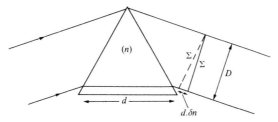

FIG. 6.16. Resolving power of a prism.

the wavefront in the new wavelength will turn through a certain angle to a position Σ' as in the figure. In order to make this angle equal to λ/D, so that it corresponds to the resolution limit, we have to make the change in optical path equal to one wavelength, i.e. we put $\delta n \cdot d = \lambda$. Thus we find

$$\frac{\delta n \cdot d}{D} = \frac{\lambda}{D}$$

or,

$$\frac{\lambda}{\delta\lambda} = d \cdot \frac{dn}{d\lambda} \tag{6.14}$$

This is the very simple formula for the spectroscopic resolving power of a prism. Strictly d must be interpreted as the difference between the extreme light paths across the beam, since the edge of the prism would not be used.

6.5. Multiple-beam interference

In Sections 6.1 and 6.2 we showed how the interference effects between two beams, giving $\cos^2$ fringes, are used. The diffraction grating (Section 6.3) works through the interference of many beams, since it is only when the phases of all these coincide (as in Fig. 6.7) that a maximum of light intensity is found. In this context the grating is a *multiple-beam interferometer*. If we illuminate a grating with collimated monochromatic light we find maxima in the far field in directions α', given by eqn (6.7), and we can plot these as in Fig. 6.17 with vertical broken lines corresponding to the order M. The angular width of each maximum is, as in Section 6.4, λ/D, where D is the full width of

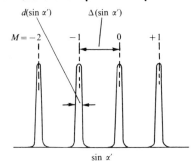

FIG. 6.17. Diffracted monochromatic light from a grating as an example of multiple-beam interference.

the diffracted beam, and we have $D = N\sigma \cos \alpha'$. Thus the angular width is

$$d\alpha' = \lambda/N\sigma \cos \alpha',$$

and the increment in $\sin \alpha'$ between successive orders is $\Delta(\sin \alpha') = \lambda/\sigma$ from eqn (6.7). Thus

$$\frac{d(\sin \alpha')}{\Delta(\sin \alpha')} = \frac{1}{N} \tag{6.15}$$

and we see that *each maximum is $1/N$th the width of the spacing between maxima*, where as before N is the number of rulings in the grating. We said that the diffraction grating can be regarded as a multiple-beam interferometer. From this point of view eqn (6.15) shows that the width of the interference maxima is found by dividing the interval between maxima by the number of interfering beams. This is a general property of multiple-beam interference effects, and it is in fact also true of two-beam interference, since two-beam fringes have equal light and dark spaces. Thus the general effect of using multiple beams is to sharpen or narrow the interference fringes. Figure 6.17 does not show in detail the distribution of light intensity between the different maxima; for the detailed calculation, see, for example, Born and Wolf (1965).

Other forms of multiple-beam interference take place between parallel reflecting surfaces. In Section 3.1 we discussed interference effects in thin films, but considered as an approximation only two beams—those first reflected from the upper and lower surfaces. These two beams would be beams 1 and 2 in Fig. 6.18, but actually the light is multiply reflected as indicated by the broken lines, and strictly all the beams should be taken into account in calculating the interference effects.

In the case of the oil film or, for example, an anti-reflection coating of magnesium fluoride on glass, the effects of the succeeding reflections are negligible, since they are much fainter than the first two, but this is not so in

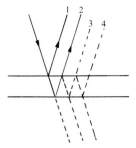

FIG. 6.18.   Multiple reflections in a thin layer.

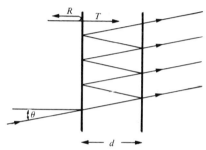

FIG. 6.19.   Principle of the Fabry–Perot interferometer. The films, in practice
supported on plates of fused silica, have high reflectivity and low transmission. The
angle of incidence is exaggerated to separate the rays, although in fact all the successive
transmitted wavefronts overlap almost completely.

other cases. Figure 6.19 shows the *Fabry–Perot interferometer*. This consists
of two accurately plane and parallel surfaces with high reflectivity and low
transmission (R and T respectively). The spacing between the layers is d and
collimated monochromatic light from a broad source falls on the first
surface.† We suppose the incident beam to be broad enough and the angle of
incidence to be small enough to ensure that many multiply reflected beams
emerge, superpose and interfere in the far-field on the right-hand side. The
interference pattern in the far-field can be brought to a focus by means of a
lens and, as for the Michelson interferometer, it must, by symmetry, consist of
concentric bright rings, each corresponding to a certain angle of incidence θ
at which all the beams of a certain wavelength are in phase.

To find the form and spacing of the fringes we note that, as in Section 3.1,
the optical path difference between successive transmitted beams is
$p = 2d \cos \theta$. Thus we can write down the complex amplitudes of the

† The reflecting surfaces could be very thin silver layers supported on glass or fused
silica. They are usually dielectric multilayers, not silver, like the end mirrors of lasers,
and typically we could have $R \sim 0.95$ and $T \sim 0.05$.

successive transmitted beams as follows, taking the origin of phase at the point of emergence of the first beam,

$$T,$$

$$TR \exp (i2\pi p/\lambda),$$

$$TR^2 \exp (i4\pi p/\lambda),$$

$$TR^3 \exp (i6\pi p/\lambda),$$

$$\cdots\cdots\cdots$$

Note that T and R refer to *intensities* but our present calculation is concerned with complex amplitudes, so that T gives the amplitude transmission through two surfaces, as required, and similarly for R. The total transmitted amplitude is the sum of terms like these. To simplify the calculation we assume the number of terms is infinite, and we then merely have to sum a geometric series with common ratio $R \exp (i2\pi p/\lambda)$. The result is

$$\text{Transmitted complex amplitude} = \frac{T}{1 - R \exp (i2\pi p/\lambda)}. \quad (6.16)$$

To get the transmitted light intensity we take the squared modulus, according to the rule in Section 1.3, giving

$$\frac{T^2}{1 + R^2 - 2R \cos (2\pi p/\lambda)} = \frac{T^2}{(1 - R)^2 + 4R \sin^2 (\pi p/\lambda)}$$

Finally, if we put $R + T = 1$, i.e. we neglect absorption in the reflecting layers, we have

Fabry-Perot transmission $I(\theta) =$

$$1 \Big/ \left\{ 1 + \frac{4R}{(1 - R)^2} \sin^2 \left(\frac{2\pi}{\lambda} d \cos \theta \right) \right\}. \quad (6.17)$$

It is easiest to see the general form of the multiple-beam fringes by regarding $I(\theta)$ as a function of the phase difference $\phi = (2\pi/\lambda)d \cos \theta$ which appears in eqn (6.17). For typical values of R and T the quantity $4R/(1 - R)^2$ is of order of magnitude 1000. Thus the denominator in eqn (6.17) is very large and the transmission is correspondingly small unless ϕ is very close to a multiple of π. The transmission function is therefore as in Fig. 6.20, i.e. the fringes have the characteristics of multiple-beam interference as described for the diffraction grating.

The Fabry–Perot interferometer can be used as a spectroscopic device and as an interferometer for the measurement of distance or of phase. In either case we have to estimate the width of the bright fringes and the spacing between fringes. A convenient way of using the Fabry–Perot for spectroscopy is to detect only the central fringe, as we described for the Michelson

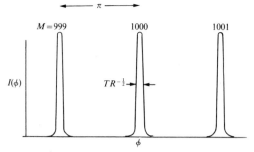

FIG. 6.20. The fringe shape for the Fabry–Perot interferometer as a function of $\phi = (2\pi/\lambda)\, d\cos\theta$, half the phase difference between the maxima. The orders of interference indicated are notional.

interferometer, and to change the wavelength by varying the spacing d. It is then called a *scanning interferometer*. In this mode the maxima for wavelength λ must occur at spacings $d = M\lambda/2$, where M is an integer, the order of interference (but in the Fabry–Perot the order may be between 10^3 and 10^6, whereas in the diffraction grating it rarely exceeds 10). The orders are numbered typically in Fig. 6.20. Let the half-width of the fringes be given by 2ε as a fraction of the order—i.e. we suppose that, if we put $d = (M + \varepsilon)\lambda/2$ in eqn (6.17), the transmission falls to 0.5. Substituting in eqn (6.17) we have

$$\frac{1}{2} = 1 \Big/ \left\{1 + \frac{4R}{(1 - R)^2} \sin^2 (M\pi + \varepsilon\pi)\right\};$$

Removing $M\pi$ from the argument of the sine and putting $\sin \varepsilon\pi \sim \varepsilon\pi$, we find

$$2\varepsilon = \frac{1 - R}{\pi R^{\frac{1}{2}}}.$$

This result is usually expressed in the form, fringe-spacing divided by fringe-width is

$$F = \frac{\pi R^{1/2}}{1 - R}. \qquad (6.18)$$

This quantity F is called the *finesse*.† We can use it to get an estimate of the resolving power of the Fabry–Perot interferometer. This is defined as for all spectroscopic devices as $\lambda/\delta\lambda$, where now $\delta\lambda$ is the measure in wavelength units of the quantity 2ε we found above. In the equation $d = M\lambda/2$ we now keep d constant and find the increment $\delta\lambda$ corresponding to an increment 2ε in M. We have

$$d = \frac{(M + 2\varepsilon)(\lambda - \delta\lambda)}{2},$$

† Equation (6.15) gave the corresponding quantity for the diffraction grating, or rather its reciprocal.

or, to the first order in small quantities,

$$\lambda/\delta\lambda = M/2\varepsilon.$$

Substituting $1/F$ for 2ε we find

$$\lambda/\delta\lambda = MF, \qquad (6.19)$$

or, *the resolving power of the Fabry–Perot is the product of the order of interference and the finesse.* By comparing eqn (6.19) with eqn (6.13) it can be seen that the finesse plays a role in the theory of the Fabry–Perot similar to the number N of interfering beams in a diffraction grating. Thus F is sometimes called the effective number of interfering beams.

In the Fabry–Perot a wavelength λ_1 may give a fringe maximum of order M at the spacing d, while at the same spacing another wavelength λ_2 may also have a maximum, but of order $M + 1$. Then we have

$$M\lambda_1 = (M + 1)\lambda_2 = 2d.$$

The interval between these two wavelengths is the maximum length of spectrum which can be studied without confusion between spectra of different orders, and it is called the *free spectral range.* Denoting this quantity by $\Delta\lambda$ we have from the above equation

$$M(\lambda + \Delta\lambda) = (M + 1)\lambda,$$

from which, if M is much greater than unity, we obtain the two useful expressions for the free spectral range

$$\Delta\lambda = \frac{\lambda}{M} = \frac{\lambda^2}{2d}. \qquad (6.20)$$

There is a similar effect of overlapping orders with the diffraction grating. Methods of avoiding confusion between overlapping orders are described in books on spectroscopic techniques.

6.6. Thin film interference devices

We mentioned antireflection coatings in Section 6.5 and we now discuss these and other related devices in more detail.

Let a surface of glass of refractive index n be coated with a layer of material of refractive index n_0, less than n, and of thickness d such that $n_0 d = \lambda_m/4$ where λ_m is a chosen wavelength, usually in the middle of the visible spectrum. Small fractions of incident light will be reflected from both surfaces, as in Fig. 6.21(a) and, since the phase changes on reflection will be the same for each fraction (Section 3.1) these fractions will be out of phase with each other by π and they will therefore interfere destructively. If furthermore $n_0^2 = n$ it is found from eqn (4.4) or (4.5) (by putting $\theta_1 = \theta_2 = 0$) that the intensities of the two reflected fractions are equal and the resultant reflected intensity will be zero.

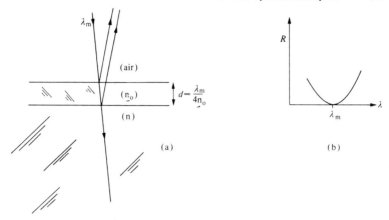

FIG. 6.21. An antireflection coating; the rays are intended to suggest plane wavefronts incident normally.

This is the principle of the simplest kind of antireflection coating. The condition $n_0 d = \lambda_m/4$ only holds at the one wavelength, λ_m, so that at neighbouring wavelengths there will be some residual reflection and the graph of reflectivity as a function of wavelength will be as in Fig. 6.21(b). We may also note that the condition of π phase difference only holds at normal incidence for λ_m; at oblique incidence eqn (3.4) gives the optical path difference between the interfering beams and eqns (4.4) and (4.5) would be used to calculate the effects separately in **p** and **s** polarized light.

The above is a simplified sketch of the theory of such coatings which ignores multiple reflections. To calculate the properties more accurately and to deal with more complicated coatings made up of several films it is necessary to use electromagnetic theory. However, it is still possible with our method to see, for example, the principle of multilayer high reflecting coatings. These are made up of several (11 or 13) layers of alternating high and low refractive index, each of optical thickness nd equal to $\lambda_m/4$, as in Fig. 6.22. Then if we apply similar reasoning to the above, recalling the rule about phase changes on reflection (Section 3.1), we find that the beams reflected from successive interfaces are successively 2π or zero behind each other in phase, i.e. they are all in phase and so a very high reflected intensity builds up. The end reflectors in lasers are made on this principle. Their behaviour at other wavelengths than λ_m and at non-normal incidence is complex and it cannot be explained by our elementary approach.

6.7. Spectroscopy in general

In the preceding sections we have described only a few of the large number of spectroscopic methods and instruments based on apparently many different

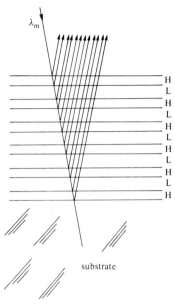

FIG. 6.22.   A multilayer coating giving high reflectivity at the design wavelength λ_M. The layers are of alternately high (H) and low (L) refractive index and the optical thickness of each is $\lambda_M/4$.

principles, but in fact there are only a few underlying ideas, and it is mainly the technical details concerned with adaptation for special purposes which differ.

Spectroscopy is concerned with measuring the proportions of different frequencies or wavelengths in a beam of polychromatic light. If we form a two-beam interference fringe system, e.g. with Young's apparatus (as in Fig. 6.1), the fringe spacing is proportional to the wavelength, and thus the intensity distribution in the fringe system formed by polychromatic light contains in an indirect or coded form the information we seek. The same is true for the Michelson interferometer—the fringe function is an encoded form of the spectrum. In both cases the decoding process is the same—taking the Fourier transform of the fringe function—but the Michelson interferometer gathers more light flux, and it is therefore preferable for most purposes.

The Michelson interferometer can also be regarded as a *multiplexer*. In telecommunications it is common practice to use a single line or channel to carry several messages simultaneously, e.g. by using the messages to modulate different carrier waves transmitted at the same time. Now if the mirror in the Michelson interferometer moves with velocity v the fringe function for monochromatic light oscillates at frequency $2v/\lambda$, as can be seen by putting $z = vt$ in eqn (6.3). Thus a wavelength λ is modulated at frequency $2v/\lambda$, and the interferometer is a multiplexing device which modulates each wavelength at a different frequency. The multiplexed signal is then decoded by taking its Fourier transform.

These ideas from communication theory have led to the development in the last 20 years of several new spectroscopic devices intended to increase light-gathering power or speed of operation. For example, we can have a grating monochromator with arrays of randomly spaced apertures in place of the normal entrance and exit slits, as in Fig. 6.23. Such a system would apparently

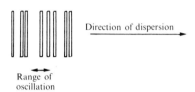

FIG. 6.23.   Multiple slits, randomly spaced, of a modulation spectrometer. The exit-slit array is similar. Wavelengths throughout the whole spectrum are transmitted, but only the wavelength for which the prism is set is strongly modulated.

have reduced resolving power corresponding to the greater wavelength range covered by the apertures, but if the prism or grating is oscillated at a certain frequency the central wavelength is strongly modulated at this frequency but neighbouring wavelengths are not.

Systems like the Michelson interferometer based on two-beam fringes always produce an output which has to be decoded to give the spectrum, and this is because the free spectral range of two-beam fringes is zero. Thus in order to produce a spectrum direct rather than encoded we have to use multiple-beam interference, e.g. a diffraction grating or a Fabry–Perot interferometer. In these and many other kinds of multiple-beam spectroscopes the bright fringes are narrow enough to allow fringes from many neighbouring wavelengths to be formed in between them.

The dispersing prism is a special case outside these classes. All the other systems we have mentioned rely on geometry alone for their effects—diffraction at slits followed by interference, or reflection at mirrors followed by interference. The prism depends on a property of a material medium, namely dispersion, whereas the other systems could operate in a vacuum with thin films of conducting, i.e. reflecting material as mirrors, screens, and beam-splitters.

In light scattered from a laser beam by, say, a colloidal suspension or a turbulent gas stream there are fluctuations in intensity which are, as explained in Chapter 1, interpreted as a spread of wavelengths in the spectrum of the scattered light. Let the intensity in the beam as a function of time be I(t); we can define the normalized autocorrelation function of the intensity as

$$C(\tau) = \frac{1}{T} \int\limits_{-\frac{1}{2}T}^{\frac{1}{2}T} I(t)I(t + \tau) \, dt$$

(see Appendix), where T is a time which is long compared to the fluctuations in question. From the autocorrelation theorem (Appendix) $C(\tau)$ is proportional to the Fourier transform of $\{G(v)\}^2$, the square of the spectrum of the light, so if we measure $C(\tau)$ we can obtain the spectrum. If the spectrum is very narrow, as it would be in scattered laser light, the delays for which $C(\tau)$ has to be measured can exceed 10^{-9} s, and then the autocorrelation can be measured directly by rapidly responding detectors which record individual photoelectrons. This technique is called *correlation spectroscopy*, and it has been developed in recent years as a method of spectroscopy suitable for very narrow spectral lines.

Problems

6.1. In Young's interference experiment the source pinhole and the receiving screen are each 1 m from the two secondary pinholes, and these are 1 mm apart. (a) What is the fringe spacing for light of wavelength 546 nm? (b) Estimate the maximum diameter of the source pinhole for fringes of good contrast to be formed. (c) What would be the effect on the fringes if the two secondary pinholes were not of equal size?

6.2. In an experiment with a stellar interferometer the fringes from the star Betelgeuse had zero visibility for a wavelength in the middle of the visible spectrum for a separation of the mirrors of 3 m. Estimate the angular subtense of Betelgeuse in arcseconds.

6.3. For the Michelson interferometer of Fig. 6.5, find an expression for the radius of the Nth circular fringe from the centre of the far-field interference pattern, in terms of f_1, f_2, and λ.

6.4   Calculate the form of the fringe function in a Michelson interferometer for a spectrum of rectangular profile. Plot a graph of this function for the case where the width of the spectrum in frequency units is 20 per cent of the mean frequency.

6.5. The red line of cadmium (644 nm), as produced by a certain discharge tube, is found to have a coherence length of 200 mm. Estimate the width of the line in wavelength and frequency units.

6.6. A reflection diffraction grating has rulings with 1 μm spacing. Draw a graph of the angle of diffraction as a function of wavelength for the first-order spectrum if the illuminating beam is at normal incidence.

6.7. In the arrangement of Problem 6.6, if the shortest wavelength to be used is 200 nm, calculate the free spectral range, i.e. the wavelength range without overlapping orders, for the first-order spectrum.

6.8. A dispersing prism has an angle of 60°, and its refractive index is 1.762 for the mercury line of wavelength 546 nm. What is the angular deviation of this wavelength at the minimum deviation setting of the prism?

6.9. The prism of Problem 6.8 has refractive index 1.791 for the mercury line of wavelength 436 nm. Estimate the spectroscopic resolving power of the prism in this region of the spectrum if it is equilateral with a 20 mm base.

6.10. Draw a graph of the fringe function of a Fabry–Perot interferometer with $R = 0.9$, $T = 0.1$. Calculate (a) the finesse and (b) the minimum transmission.

6.11. What is the spectroscopic resolving power of the Fabry–Perot in Problem 6.10 if the spacing of the plates is 5 mm? Calculate its free spectral range at wavelength 500 nm.

7. Laser light

Glass flashing. That's how that wise man what's his name with the burning glass. Then the heather goes on fire.

James Joyce: Ulysses

For the purposes of this book lasers are simply sources of very intense, spectrally pure, and spatially coherent light. Everything we describe could, in principle, be done with ordinary thermal light passed through a monochromator of very high resolving power and then through a diffraction-limited pinhole. However, the light intensity would be so low that the experiments would in practice be impossible. Helium–neon lasers of the kind now commoner than sodium lamps in many laboratories produce about 1 mW of coherent monochromatic light of wavelength 632.8 nm, but the brightest available thermal source, an ultra-high pressure mercury lamp, would produce about 10 orders of magnitude less light power of the same coherence and monochromaticity.

7.1. Laser beams

The formation of the characteristic narrow intense beam of a helium–neon laser can only be fully explained by a detailed treatment such as is given by, e.g. Svelto (1976), but we can suggest in a general way how the beam shape occurs. That the laser light is highly coherent is easily confirmed experimentally, since interference fringes of high contrast can be formed from laser light with none of the usual precautions which have to be taken when ordinary thermal sources are used. Thus from our discussions of coherence (Sections 1.5 and 3.5) the laser beam must be a monochromatic wavetrain behaving as if it originated from a single point source, i.e. at any given point in the beam there is a single unique wave vector. This is called a single mode and it is like the single modes which can be generated in alternating current circuits, mechanical oscillating systems, acoustic oscillators such as organ pipes, etc. In each case a resonant system is needed in which to generate the oscillation and for the HeNe laser the resonant system or cavity is formed by two mirrors at either end of a discharge tube containing a mixture of He and Ne. The form of the mode, i.e. the shape of the laser light beam, is found by asking what distribution of complex amplitude across one of the mirrors will, on propagating to the other mirror and back again, reproduce itself so that a stable mode structure can be generated by the electrical discharge in the tube. The question may be rephrased, what distribution of complex amplitude will reproduce its shape on propagating to the far field, or in the terms of Section 3.4, what function is its own Fourier transform? There are many solutions to this mathematical problem but it turns out that the simplest solution is the one which is appropriate to the laser under ordinary conditions. This is the Gaussian function, of which the modulus of the complex amplitude has the form $\exp - \pi(x^2 + y^2)$, and it can be shown (Svelto 1976) that as the beam propagates the intensity profile always has the same shape, merely expanding or contracting laterally. This is shown in Fig. 7.1 where the arrowed lines indicate levels of constant intensity in the beam. After focusing by a lens the

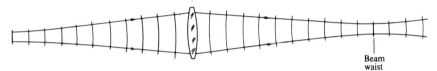

Beam
waist

FIG. 7.1.   A Gaussian beam. The arrowed lines connect points on the phase fronts where the intensity is $1/e^2$ that at the centre of the beam. The phase fronts are plane at a beam waist.

beam contracts to a 'waist', which is like the focus of an ordinary beam from a 'point source' of thermal light, and expands again. The phase fronts are shown in the diagram; at some distance from the waist they are spherical surfaces centred almost exactly on the waist, i.e. they are like geometrical wavefronts. Near the waist the phase fronts decrease in curvature and they are plane at the waist. We remarked in Section 2.2 that geometrical wavefronts do not coincide with physical wave fronts near foci and Fig. 7.1 shows this clearly, since the geometrical wavefronts would shrink to a point at the focus, i.e. the beam waist. In fact the beams discussed in Chapters 2 and 3 have uniform intensity across the wavefronts until they are cut off sharply by an aperture stop or the rim of a lens and for such beams the form of the phase fronts near the focus is more complicated than in Fig. 7.1 (see Born and Wolf 1975).

7.2. Coherent light speckle

If an optically rough or scattery surface is illuminated with laser light a striking effect is seen: the surface is covered with a fine network of bright and dark patches. This effect has become known as *laser speckle*, although it can also be observed with thermal light of sufficient coherence. The same effect is seen in the far-field pattern from a rough surface, e.g. ground glass, illuminated with a laser beam as in Fig. 7.2. The explanation is that each point on the surface scatters a beam which is coherent with the beams from all the other scattering points, but there are random phase relationships between these beams, so that in the far-field we see a superposition of interference patterns between all pairs of scattering points. These interference patterns have random spatial frequencies, phases, directions, and contrasts, and the coherent sum of them all is the speckle pattern. The maximum spatial frequency in the pattern corresponds to interference between pairs of scattering points at opposite ends of a diameter of the ground glass.

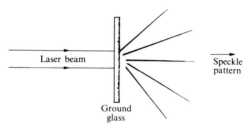

FIG. 7.2.   Formation of a speckle pattern by a random diffuser. Speckle is formed at all distances from the diffuser, and the pattern has nearly the same statistical properties at all distances.

The explanation of the speckle pattern seen by looking at the surface of the scatterer rather than at the far-field is slightly different. In looking at a certain point on the surface we see the coherent sum of all the scattered beams within the radius of a resolution limit around that point, and this coherent sum may again have a range of intensities, depending on the random phases of the scattered beams. This argument shows that the scale of detail seen in the speckle pattern on a scattering surface corresponds precisely to the resolution limit of the optical system used to view it.

Coherent light speckle obscures detail in the structure of the image of an object formed in coherent light and it is thus a great nuisance in image-forming systems which use coherent light. On the other hand, since it is an interference pattern, it may carry information about the statistics of the scattering surface, i.e. the height variations and their lateral scale, and also about movements of the surface; this is being turned to account in some recent applications.

7.3. Holography

In Section 3.1 we saw how two intersecting coherent collimated beams produce straight and parallel interference fringes. If the beams intersect at an angle θ the fringe spacing perpendicular to the bisector of the angle θ is $\sigma = \lambda/(2 \sin \frac{1}{2}\theta)$. If we record these fringes on a photographic plate or other recording medium by placing it in the beam as in Fig. 7.3 and then developing the plate, we have, in effect, a diffraction grating (Chapter 6). We next set up the grating with one of the beams switched off, say beam 2, as in Fig. 7.4, and we find several diffracted beams of different orders. The grating equation (eqn (6.7)) for a transmission grating can be written

$$\sin \alpha - \sin \alpha' = M\lambda/\sigma, \tag{7.1}$$

and in this case we have $\alpha = \theta/2$. Then for the zero-order diffracted beam ($M = 0$) we have $\alpha' = \alpha$, i.e. an undeviated beam, and for $M = 1$ we find

$$\sin \alpha' = \sin \theta/2 - \lambda/\sigma;$$

but from the way we made the grating we have $\lambda/\sigma = 2\sin(\theta/2)$, so that for the first-order beam,

$$\alpha' = -\theta/2. \tag{7.2}$$

This means that the first-order diffracted beam travels in the direction in which beam 2 of Fig. 7.3 was travelling. Thus we have 'reconstructed' beam 2 by illuminating the grating with beam 1.

Another way of looking at this process is to say that in photographing the fringe pattern we are attempting to record the complex amplitude of beam 2

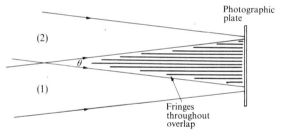

FIG. 7.3.   Two collimated coherent beams form fringes throughout the overlap region, with spacing $\frac{1}{2}\lambda/\sin\frac{1}{2}\theta$. The plate records a grating pattern with this spacing.

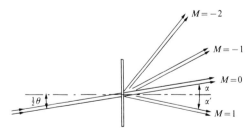

FIG. 7.4.   Diffracted beams from the sinusoidal grating produced as in Fig. 7.3. The first-order beam is the holographic reconstruction.

at the plane of the photographic plate. The photograph does not record unambiguously everything about this complex amplitude distribution. Thus the fringe spacing tells us that the phase changes by 2π every fringe, but we do not know in which direction the phase is increasing. This ambiguity can be regarded as the cause of the appearance of the other diffracted beams of orders 0, -1, 2, 3, etc. On the other hand, if we had attempted to record beam 2 by placing the photographic plate in it without beam 1 to form fringes, we should have obtained merely a uniform blackening, i.e. a record of the intensity, and this could not contain any information about the direction from which beam 2 had come.

Beams 1 and 2 can be regarded as originating from point sources P_1 and P_2 at infinity. Then in the terminology of holography we formed a *hologram* of P_2 with P_1 as reference, and we then reconstructed P_2 with the same reference point and using the same wavelength light.

A similar thing occurs if P_1 and P_2 are at finite distances, as in Fig. 7.5. The fringes on the hologram are now curved, but it is again found that if the hologram is illuminated with P_1 alone then P_2 will be reconstructed (as in Fig. 7.6), and vice versa.

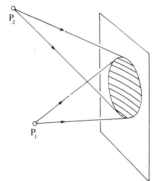

FIG. 7.5.   Formation of a hologram of a point P_2 with reference point.

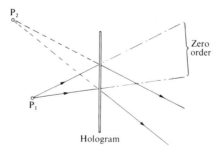

FIG. 7.6.   Reconstruction of the hologram of Fig. 7.5. The first-order diffracted beam appears to come from the point P_2 behind the hologram.

Now suppose we have a reference point P_1 and an array of N object points $P_2, P_3, \ldots P_{N+1}$, and we record the interference pattern as before. The sets of fringes formed by interference between P_1 and each of the other points will all be formed, and, provided the recording medium has enough range, they will be superposed. Then on reconstructing with P_1 the image of the array of N points will be formed. This is the principle of holography as invented by D. Gabor in 1948. Ultimately the array $\{P_i\}$ becomes a continuous distribution, and the hologram is a record of the interference pattern between the reference beam and the complex amplitude scattered from this continuous distribution. Let $E_0(x, y)$ be this complex amplitude as a function of coordinates (x, y) in the plane of the hologram plate, and let $E_r(x, y)$ be the complex amplitude due to the reference beam. The light intensity in the interference pattern is $|E_0 + E_r|^2$, and to a reasonable approximation the complex amplitude transmission T_a of the developed hologram is a linear function of this,

$$T_a = k_0 - k_1 |E_0 + E_r|^2,$$

where k_0 and k_1 are constants. In the reconstruction process the complex amplitude transmitted by the hologram is $T_a E_r$, and on multiplying out the squared modulus we see that this is

$$T_a E_r = k_0 E_r - k_1 E_r E_0 E_0^* - k_1 E_r^2 E_0^*$$
$$- k_1 |E_r|^2 E_0 - k_1 E_r |E_r|^2. \tag{7.3}$$

The intensity $|E_1|^2$ of the reference beam is roughly constant over the hologram, since it comes from a point source at some distance. Thus the fourth term in eqn (7.3) is equal to a constant multiplied by the complex amplitude $E_0(x, y)$ at the hologram due to the original object, and this term therefore accounts for the reconstructed image of the object. There are, however, four other terms in eqn (7.3) to be accounted for. It is easier to see the significance of these by assuming that the reference beam is collimated and that it meets the hologram plate at an angle of incidence θ, as in Fig. 7.7. Then the complex amplitude in the reconstructing beam, assumed to be the same as in the reference beam, can be taken as $\exp\{i(2\pi/\lambda)y \sin \theta\}$. Substituting this value in eqn (7.3) we obtain for the transmitted amplitude on reconstruction, after regrouping the terms,

$$T_a E_r = \{k_0 - k_1(1 + |E_0|^2)\} \exp\{i(2\pi/\lambda)y \sin \theta\}$$
$$- k_1 E_0^* \exp\{2i(2\pi/\lambda)y \sin \theta\} - k_1 E_0. \tag{7.4}$$

Of these terms the first represents a wave travelling in the same direction as the reconstructing beam but with different intensity; it corresponds to the zero-order diffracted beam in Fig. 7.4 and we show it as beam 1 in Fig. 7.8. The second term, on account of the exponent $2i(2\pi/\lambda) \sin \theta$, travels in a general direction θ' such that $\sin \theta' = 2 \sin \theta$. It can be shown to produce a

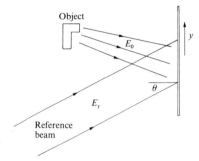

FIG. 7.7. Forming a hologram of an extended object with an oblique reference beam.

spurious real image as in Fig. 7.8, and it corresponds to the diffracted beam of order -1 in Fig. 7.4. The last term produces the correct virtual-image reconstruction of the object. The detailed properties of the photographic emulsion or other recording medium control the relative intensities of these various images.

Holographic images have the remarkable property of being three-dimensional to the extent that the method of illumination of the object and the angular range of viewing through the hologram permit. This can be seen from Fig. 7.9, which shows a typical arrangement of apparatus for holography.

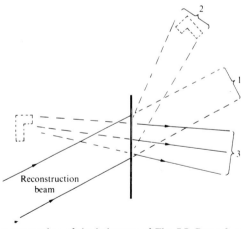

FIG. 7.8. Reconstruction of the hologram of Fig. 7.7. Beam 2, produced by the second term in eqn (7.4), produces another image of the object. This image is real but usually very distorted. Beams 1, 2, and 3 correspond to the three terms of eqn (7.4).

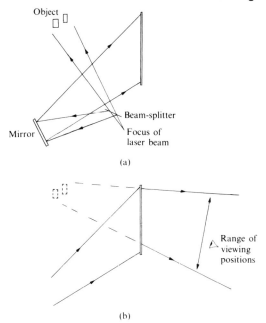

Object

Beam-splitter

Mirror

Focus of
laser beam

(a)

Range of
viewing
positions

(b)

FIG. 7.9. (a) A typical optical arrangement for taking a hologram. The laser beam is focused down by means of a microscope objective and allowed to diverge. (b) Reconstruction, showing that the image can be seen from a range of angles to give stereoscopy.

7.4. Hologram interferometry

The holographic image is a reconstruction of the complex amplitude of the light scattered from the object with a particular geometry of illumination. Suppose that a hologram of an object is taken but that the object is left in position and with the same illumination at the reconstruction stage. If the hologram plate is replaced exactly in its original position after development it will produce a virtual image exactly coinciding with the object. In practice there will generally be a slight displacement between them, as in Fig. 7.10. The light coming through the hologram scattered from the actual object is coherent with the light from the reconstructed, virtual object, and there can therefore be interference between these two beams. If the relative displacement is small only one object will be seen but its surface will appear to be covered by interference fringes which indicate the relative displacement between the original object and its reconstructed image. This is the principle of hologram interferometry. It can be done with optically rough surfaces, in fact it is best done with rough surfaces, since then the illumination and

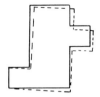

FIG. 7.10.   A reconstructed image (broken line) superimposed on the slightly displaced object. The displacement appears as a fringe pattern in hologram interferometry.

viewing conditions are less critical. Classical interferometry, on the other hand, can only be done with smooth, mirror-like surfaces.

Hologram interferometry is used in engineering and metrology for determining displacements and strains of surfaces in a variety of applications. The mode of formation of the fringes is intrinsically more complicated than in classical interferometry, as can be seen from Fig. 7.11. Each of the two

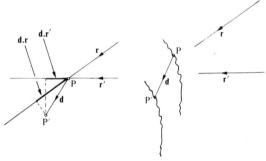

Fig. 7.11.   The optical path difference mapped by the fringes in hologram interferometry. The local displacement vector is $d(=PP')$, and the local illumination and viewing directions are specified by the unit vectors r and r'.

surfaces, that of the original object and its reconstructed image, appears to the observer to be covered by a speckle pattern (Section 7.2), and this is itself an interference pattern. The 'fringes' seen in hologram interferometry are due to interference between these two speckle patterns. The speckle patterns themselves are congruent but displaced, and therefore the displacement measured by the fringes is that corresponding to the displacement of a definite point, e.g. P, to P' in Fig. 7.11. This displacement is in general not normal to the surface, and it may even be in the plane of the surface, but the fringes will still map it as an optical path difference.

Let r and r' be unit vectors along the directions of illumination and viewing near P and let d be the vector displacement from P to P'. Then we see from the figure that the total path difference indicated by the fringe system near P is

$$W = \mathbf{d} \cdot (\mathbf{r} + \mathbf{r}'), \qquad (7.5)$$

which is a rather complicated function of the directions $\mathbf{r}$ and $\mathbf{r}'$ and the displacement. For example, we can measure a displacement of a surface in its own plane, as in Fig. 7.12, if the illuminating and viewing directions are suitably chosen, whereas such a measurement would give a null result in classical interferometry with smooth surfaces.

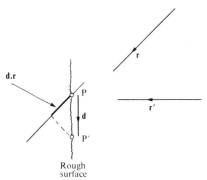

FIG. 7.12.   Measurement of in-plane displacement by hologram interferometry. In the arrangement shown the surface is viewed normally, although this is not essential. The fringes map the quantity $\mathbf{d} \cdot \mathbf{r}$, which, as can be seen, is non-zero for an in-plane displacement.

For example, let the in-plane displacement referred to in Fig. 7.12 be a small rotation $\delta\phi$ of a disc of metal in its own plane about its centre. To find the form of the resulting fringes we have from Fig. 7.13a

$$\mathbf{d} = (\rho \cos \phi \delta\phi, \rho \sin \phi \delta\phi, 0)$$

and we can take $\mathbf{r}$ as given by

$$\mathbf{r} = (0, \sin \theta, \cos \theta)$$

the z-axis being perpendicular to the plane of the disc. Then if as in Fig. 7.12 we view normally to the disc we have from eqn (7.5) for the path difference to be mapped by the fringes,

$$W = \rho \sin \phi \sin \theta \delta\phi \qquad (7.6)$$

The fringes are loci of constant W so that if the illumination vector $\mathbf{r}$ is constant over the disc eqn (7.6) shows that the fringes are loci of constant $\rho \sin \phi$, i.e. they are straight, equidistant and parallel to the y axis, as in Fig. 7.13b.

The form of eqn (7.5) suggests that the same fringe pattern could be given

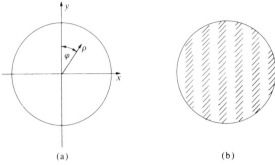

FIG. 7.13.   Hologram interference fringes from an in-plane displacement; (a) coordinates in a disc rotated a small angle $\delta\phi$ in its own plane, (b) the fringes.

by different combinations of displacements and illuminating and viewing conditions since W depends on three independent vectors, all of which could vary over the surface being examined. Thus in our example of Fig. 7.13 we could get the same fringe pattern if, say, the rotation $\delta\phi$ were doubled and $\sin\theta$ (for the illumination vector) were halved. Again, the same fringe pattern would result if, with the same illuminating and viewing conditions, the displacement were changed from an in-plane rotation to a small wedge, i.e. if **d** were given by

$$\mathbf{d} = (0, 0, \varepsilon x)$$

where $\varepsilon = \delta\phi \tan\theta$. This indeterminacy, which is not a feature of classical interferometry with specular surfaces, can in practice be overcome by taking at least two interferograms with different illumination or viewing angles, or sometimes by making use of other information about the kind of strain or displacement to be expected.

Hologram interferometry can be done in different ways. A variation of the method described above is to take two holograms on the same plate, one before and one after the displacement. The object is then removed and fringes are formed between the two reconstructed images; this is called the frozen fringe method. In another variation a hologram is taken of a vibrating surface with a time exposure lasting for many periods of the vibration. The reconstructed image carries fringes showing the form and amplitude of the mode of vibration.

7.5. Holographic diffraction gratings

We saw in Section 7.3 that the photographed image of the fringes formed between two collimated beams forms a diffraction grating. This process can be carried out with a recording medium such as a photo-resist which forms a relief structure indicating the fringe distribution (as in Fig. 7.14), and after

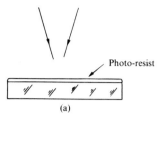

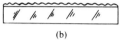

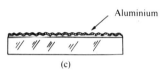

FIG. 7.14. Making a holographic diffraction grating. (a) A glass blank coated with photo-resist is illuminated with crossing laser beams to form the fringes. (b) The photo-resist is developed to give a contoured surface. (c) An aluminium reflecting coating is applied.

aluminizing the resist a reflecting diffraction grating is formed. Techniques are available for shaping the groove profile to produce a blaze, i.e. to direct most of the light into a single order of diffraction. Also the method can be applied to gratings on curved surfaces, and by careful choice of geometry better image formation can be obtained than in conventionally ruled gratings.

Hologram interferometry and the manufacture of diffraction gratings are probably the most important practical applications of holography at present.

7.6. Spatial filtering

We saw in Section 3.4 that the complex-amplitude distribution in the far-field diffraction pattern of an aperture with complex amplitude variations in it is the Fourier transform of these complex amplitude variations to a suitably chosen scale. This could be realized experimentally as in Fig. 7.15. A transparency placed in an aperture at the front focal plane of a lens is illuminated with collimated coherent light. The far-field diffraction pattern of

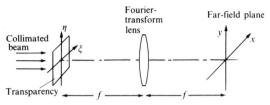

FIG. 7.15.   Optical transforms. The complex amplitude in the farfield is the Fourier transform of the complex amplitude at the object transparency.

the aperture and transparency is formed at the other focal plane, and from Chapter 3 the complex amplitude in this plane is

$$E(x, y) = \iint\limits_{-\infty}^{\infty} T_a(\xi, \eta) \exp -\left\{\frac{2\pi i}{\lambda f} (x\xi + y\eta)\right\} d\xi \, d\eta,$$

or, putting

$$x/\lambda f = s, \; y/\lambda f = t,$$

$$E(s, t) = \iint\limits_{-\infty}^{\infty} T_a(\xi, \eta) \exp - 2\pi i(\xi s + \eta t) \, d\xi \, d\eta. \tag{7.6}$$

Here s and t are spatial frequency components of the complex amplitude transmission $T_a(\xi, \eta)$, just as we defined spatial frequency components of an intensity distribution in Section 5.5. Thus the complex amplitude at (x, y) in the Fourier plane is proportional to the amount of complex amplitude with spatial frequency components $(x/\lambda f, y/\lambda f)$ in the original transparency. To take a simple example, suppose the transparency is a sinusoidal amplitude grating of complex amplitude transmission

$$T_a = 1 + \cos 2\pi s_0 \xi.$$

If we write this in the form

$$T_a = 1 + \tfrac{1}{2} \exp (2\pi i s_0 \xi) + \tfrac{1}{2} \exp (-2\pi i s_0 \xi),$$

we see (Appendix) that the Fourier transform consists of delta functions of magnitudes $1, \tfrac{1}{2}$, and $\tfrac{1}{2}$ at the origin and at $(\pm s_0, 0)$. Thus in the Fourier plane we should see bright spots P_0, P_{-1}, P_1, at the centre and at $(\pm \lambda f s_0, 0)$, as in Fig. 7.16. From the point of view of diffraction grating theory these are merely the spectra of order zero and ± 1, and it would be possible to develop the theory of diffraction gratings along these lines.

Now suppose we add to the optical system of Fig. 7.15 another lens of focal length f, as in Fig. 7.17. By elementary geometrical optics we expect to find, at the second focal plane of this lens, an image of the original object, since this

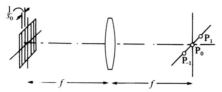

FIG. 7.16. The transform of a grating with amplitude transmission of sinusoidal form. If the grating spatial frequency is s_0 the $+1$ and -1 orders are $\lambda f s_0$ from the zero order in the transform plane.

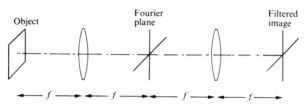

FIG. 7.17. A spatial filtering apparatus. The filters are placed in the Fourier or far-field plane, and the filtered image appears at the right.

object is imaged at infinity in the intermediate space; this is found to be broadly correct. However, if we were to obstruct part of the Fourier plane we should be removing some of the spatial frequency components, and the image would appear changed accordingly. In the above example, the intensity distribution in the object transparency is $|T_a|^2$, i.e.

$$I_0(\xi) = 1 + 2\cos 2\pi s_0 \xi + \cos^2 2\pi s_0 \xi$$

$$= \tfrac{3}{2} + 2\cos 2\pi s_0 \xi + \tfrac{1}{2}\cos 4\pi s_0 \xi, \qquad (7.8)$$

and the image would be the same. Now suppose that we put a filter or mask in the Fourier plane which removes the zero-order component, i.e. the bright spot P_0. The complex amplitude in the image will then be

$$T_a' = \cos 2\pi s_0 \xi,$$

and the identity will be the squared modulus of this,

$$I_a'(\xi) = \tfrac{1}{2} + \tfrac{1}{2}\cos 4\pi s_0 \xi. \qquad (7.9)$$

Comparing eqns (7.8) and (7.9) we see that there has been a complete change in the appearance of the image. Remembering that what we see or detect is intensity, not complex amplitude, eqn (7.8) represents a periodic structure with basic spatial frequency s_0 (but also with a harmonic of frequency $2s_0$), whereas in eqn (7.9) the basic frequency is $2s_0$. Figure 7.18 shows to scale the relative intensities represented by eqns (7.8) and (7.9).

Ernst Abbe explained these effects physically 100 years ago in developing

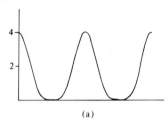

(a)

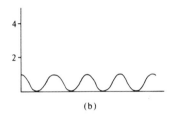

(b)

FIG. 7.18.   Images in coherent illumination: (a) the light intensity in the image of a sinusoidal complex amplitude distribution, as given in eqn (7.8), (b) the intensity in the image of the same object but with the zero spatial frequency component of complex amplitude removed.

the theory of the resolving power of the microscope. Abbe suggested that the image formed in the final image plane could be regarded as an interference pattern between sources in the Fourier plane. Thus in our example the three spots in the Fourier plane in Fig. 7.16 produce intersecting coherent plane waves in the final image space, and the image is the interference pattern between these plane waves. This pattern will have a basic periodicity corresponding to the angle between the waves from P_{-1} and P_0 (or P_0 and P_1). If P_0 is removed the interference pattern changes and its basic frequency is doubled, since it corresponds to the angle between the waves from P_{-1} and P_1.

Abbe also pointed out that if P_{-1} and P_1 are removed there will be no interference, merely uniform illumination in the image plane from a single point P_0. If we regard the two lenses as forming an imaging system, this shows why a certain minimum aperture is necessary to resolve a given spatial frequency. Specifically, the angular subtense of the semi-aperture must be at least P_0P_1/f; i.e. the collecting semi-angle α of the first lens must satisfy

$$\sin \alpha \geq \lambda s_0.$$

Thus the minimum resolvable separation $1/s_0$ is

$$1/s_0 = \lambda \sin \alpha \qquad (7.10)$$

for spatial frequencies in complex amplitude.† This may be compared with eqn (5.2), which refers to incoherent illumination. The functional dependence on wavelength and aperture is the same, but the proportionality constant is different.

Returning to spatial filtering, it is possible to put at the Fourier plane a filter to change either the phase or the amplitude of certain frequency components in any desired way. As a simple example, the dot structure in half-tone pictures may be filtered out in this way. If a transparency of, for example, an aerial photograph has prominent features running parallel to a certain direction, these may be filtered out by means of an opaque strip across the Fourier plane at right-angles to the direction. This device may make it easier to see other features.

The re-imaging stage of Fig. 7.17 can be regarded as taking the inverse Fourier transform of the complex amplitude distribution in the Fourier plane. Thus from eqn (7.6) the final image is the original object,

$$T_a'(\xi, \eta) = \int\int\limits_{-\infty}^{\infty} E(s, t) \exp\{2\pi i(s\xi + t\eta)\} \, ds \, dt. \qquad (7.11)$$

If we differentiate this relationship with respect to ξ we obtain

$$\frac{\partial T_a(\xi, \eta)}{\partial \xi} = \int\int\limits_{-\infty}^{\infty} 2\pi i s E(s, t) \exp\{2\pi i(s\xi + t\eta)\} \, ds \, dt, \qquad (7.12)$$

so that the derivative of the original is obtained by putting a filter in the Fourier plane with complex-amplitude transmission $2\pi i s$ or, changing back

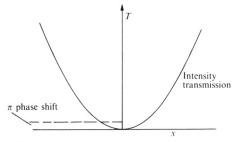

FIG. 7.19.   A differentiating filter.

to the actual coordinate in the Fourier plane, $2\pi i x/\lambda f$. This is a linear variation of amplitude transmission across the x-direction.

The factor i is not important, since it implies a constant phase shift over the whole plane, but since x changes sign on crossing the axis this implies a phase shift of π. Thus the filter would be as in Fig. 7.19 with a quadratic intensity

† Note, however, that because of the process of taking the squared modulus, a given spatial frequency in complex amplitude may be doubled in intensity.

transmission across the x-direction and a $\lambda/2$ film across one half. Many other devices of this kind are possible for analogue computing in the Fourier domain.

Problems

7.1. A piece of ground glass of diameter D is illuminated uniformly by a laser beam, and a speckle pattern is formed on a screen at a distance L. Show that the smallest detail in the pattern is of order of size $\lambda L/D$. If the screen is viewed from 250 mm and if L is 2 m, how large must the ground glass be to make the smallest detail unobservable to the eye?

7.2. In a holography experiment the reference source and the object are both 1 m from the hologram plate, and they are 100 mm apart. Estimate the scale of detail in the hologram fringes if the wavelength is 632.8 nm.

7.3. A collimated reference beam is used to form a hologram of an illuminated pinhole. Sketch arrangement and discuss the form of the hologram fringes.

7.4. In a hologram interference experiment with a helium-neon laser the surface under test is illuminated and viewed at normal incidence. If the displacement to be determined is 2500 nm, how many fringes will it be represented by (a) if the displacement is normal to the surface; (b) if it is at 45° to the normal; and (c) if it is in the plane of the surface? In case (c) suggest a way of improving the sensitivity of the technique.

7.5. A plane diffraction grating is to be produced holographically using light of wavelength 632.8 nm. Sketch the arrangement to be used, and calculate the required angles if the grating is to have 1000 rulings per millimetre. What is the closest grating spacing which could be made in this way?

8. Optical light guides

Draw out a glass rod to a thin fibre in a bunsen burner flame and break the ends of the fibre from the thicker parts. If one end is applied to a light source such as a lamp bulb or a laser the light is conducted down the fibre and it shines out brightly at the other end. This simple experiment illustrates the basis of two modern technologies. In this chapter we shall apply the ideas developed in the earlier parts of this book to such light guides.

8.1. The acceptance angle of a light guide

The mechanism of transmission of light along a guide is, in the geometrical optics approximation, total internal reflection as described in Section 2.2. Figure 8.1 shows a cylinder of material of refractive index n; if a ray of light

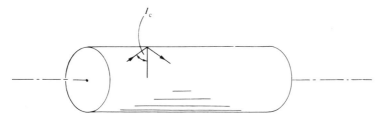

FIG. 8.1.   A solid light guide of cylindrical cross-section; the light is confined to the guide by total internal reflection.

travelling in a plane through the axis is incident on the wall at an angle greater than the critical angle I_c given by $\sin I_c = 1/n$ it will be totally internally reflected; it will carry on in this way if the guide is straight until it reaches the end of the guide. Rays in planes through the axis are called meridian rays; *skew* rays do not lie in an axial plane but equally it can be seen that a skew ray once reflected at an angle greater than the critical angle will carry on in a path with successive points of incidence lying on a helix on the surface of the guide, as in Fig. 8.2.

It is essential to have a very clean perfectly polished surface without

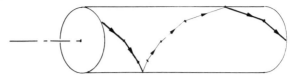

FIG. 8.2.   A skew ray propagated inside a light guide.

scratches on glass to achieve total internal reflection without losses. This is
not a very serious worry with prisms such as that shown in Fig. 2.5, since the
light is only reflected once and small losses do not matter, but in a long light
guide a given ray may be reflected very many times, so that losses due to
surface defects must be minimized; the experiment described at the beginning
of the chapter will illustrate this because specks of dust and scratches on the
fibre will appear bright when light is transmitted through it. Such surface
losses are prevented by a protective coating or *cladding*, as in Fig. 8.3, of

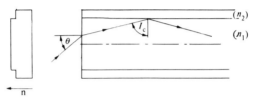

FIG. 8.3.   A protective cladding for a light guide; a meridian ray is totally internally
reflected at the interface between the guide and the cladding. The diagram on the left
indicates the refractive index variation across a diameter.

refractive index say n_2 which is lower than the index n_1 of the guide. Then
there will be a critical angle I_c given by

$$\sin I_c = n_2/n_1 \qquad (8.1)$$

for total internal reflection at the interface and damage to the outer surface of
the protective layer will not matter.

Figure 8.3 can be used to calculate the *acceptance angle* of the light guide.
This is the maximum angle of incidence θ on a square cut end face for which
any ray is totally internally reflected at the interface and therefore transmitted
without loss. The diagram shows a meridian ray and it can be seen that the
angle of refraction of this ray on entering the end face is $\pi/2 - I_c$. Thus on
applying Snell's law (Section 2.2) we find for the sine of the acceptance
angle θ,

$$\sin \theta = \sqrt{(n_1^2 - n_2^2)}. \qquad (8.2)$$

This quantity is called the numerical aperture of the light guide, by analogy
with the use of the same term for image-forming systems such as the

microscope (Section 5.4). The above argument was developed for a meridian ray; it can be seen that any skew ray incident on the end face at an angle less than θ will be transmitted. Thus all rays entering inside a solid angle $\pi \sin^2 \theta$ will be transmitted and the numerical aperture can be taken as a measure of the capacity of the guide to transmit light power. For example for a core refractive index of 1.52 the cladding index might be 1.48, giving a numerical aperture 0.35, or an acceptance angle θ of 20°.

At this point some reservations should be made. First, it can easily be seen that although θ as given by eqn (8.2) is the limiting angle for meridian rays, yet skew rays incident at greater angles to the end face can be transmitted, depending on their distance of closest approach to the axis of the fibre. Thus the guide could transmit more light flux than might be supposed from the value of the numerical aperture. Secondly our argument has been in terms of geometrical optics, and while this is accurate enought to predict the main properties of guides with lateral dimensions which are large compared to the wavelength of the light it will not serve for smaller guides; for these a treatment based on electromagnetic theory would be needed (see, e.g. Midwinter 1979).

Returning now to the properties of the guide we see that if the image of a light source is focused onto the end of the guide as in Fig. 8.4, so that it

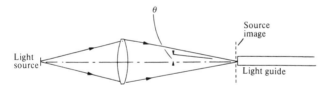

FIG. 8.4.   Illuminating a light guide. The image of the source should be at least as large as the end of the guide and the convergence angle of the illuminating beam should be at least as great as the acceptance angle of the guide.

illuminates all of the end, and if the convergence angle of the light from the lens is greater than θ, the acceptance angle of the guide, then the guide will accept as much light flux from the source as it can. Let the radiance of the source (Section 2.6) be B W mm^{-2} sr^{-1}, and let it be a Lambertian radiator; this means that the flux emitted in a direction making an angle ϕ with the surface of the source is proportional to $\cos \phi$; many thermal sources radiate approximately in this way but not, of course, lasers. If the diameter of the core of the guide is $2a$ then the flux collected by the guide is, neglecting the extra skew rays mentioned above, and neglecting also reflection losses at the entry face

$$\pi^2 a^2 B \sin^2 \theta \text{ W};\tag{8.3}$$

this expression is proved in Problem 8.2.

8.2. Graded index guides

The clad guides described in Section 8.1 can be of any diameter, depending on the application, and in fact some are made as rigid rods several millimetres in diameter. For optical communication purposes (see Section 8.4) there are advantages in having very small effective diameters, i.e. comparable with the wavelength of the light, and these have a different structure from the guides described in Section 8.1. The *graded index* guide is a cylinder in which the refractive index decreases smoothly from the axis radially outwards. The refractive index distribution may be represented, as in Fig. 8.5, by an equation such as

$$n = n_0 - \alpha r^2 \qquad (r < r_0)$$
$$n = n_0 - \alpha r_0^2 \qquad (r > r_0)$$

(8.4)

The geometrical optics of a graded index fibre of this kind is more complicated than for the simple fibre of Fig. 8.3; these latter may be called, correspondingly, step index fibres. However, if we consider rays in a meridian or axial plane we can see how a guiding effect is obtained. Consider in Fig. 8.6

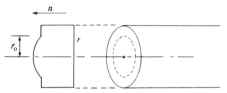

FIG. 8.5.   Refractive index profile of a graded index light guide.

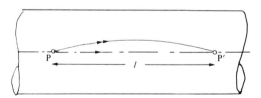

FIG. 8.6.   Focusing effect of a graded index guide.

an axial segment PP' of length l say; the optical path length from P to P' along the axis is $n_0 l$, from eqn (8.4). Now consider a second path from P to P' as indicated by the double arrows; the geometrical length of this path is greater than l but from eqn (8.4) it passes through a region of lower refractive index than the axial path, so it is possible that the decrease in index combined with the increase in geometrical distance may make the optical path length from P to P' the same along the double arrowed ray as for the axial path. Thus as in Section 2.3, we can invoke Fermat's principle and claim that P and P' could

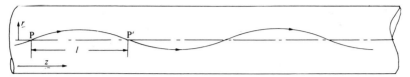

FIG. 8.7. Path of a meridian ray in a graded index guide.

be object and image, since the optical path length from P to P' is constant for all rays within a certain angle. We can quantify this as follows.

According to our reasoning above a meridian ray should follow a periodic path as in Fig. 8.7. Let the length of each half cycle be l; then we can represent the path by

$$r = a \sin (\pi z/l), \qquad (8.5)$$

where z is the axial coordinate and a is a constant for the particular ray chosen. An element of length ds along this ray is

$$ds = \sqrt{(dz^2 + dr^2)}$$

$$= \left\{ 1 + \left(\frac{\pi a}{l} \cos \frac{\pi z}{l} \right)^2 \right\}^{1/2} dz.$$

Thus the optical path length along the ray is, from Section 2.2,

$$\int n\, ds = \int_0^l (n_0 - \alpha r^2) \left\{ 1 + \left(\frac{\pi a}{l} \cos \frac{\pi z}{l} \right)^2 \right\}^{1/2} dz. \qquad (8.6)$$

Now let a be small so that only rays close to the axis, i.e. paraxial rays, are considered; specifically, we neglect powers of a above the square. Then we can substitute for r in eqn (8.6) from eqn (8.5), expand the square root and simplify to obtain

$$\int n\, ds = \int_0^l \left(n_0 - \alpha a^2 \sin^2 \frac{\pi z}{l} \right) \left\{ 1 + \tfrac{1}{2} \left(\frac{\pi a}{l} \right)^2 \cos^2 \frac{\pi z}{l} \right\} dz.$$

Then using $\cos 2\theta = 2 \cos^2 \theta - 1 = 1 - 2 \sin^2 \theta$ we finally obtain, on again neglecting a^4, etc.

$$\int n\, ds = \int_0^l \left(n_0 + a^2 \left\{ \frac{n_0 \pi^2}{4l^2} - \frac{\alpha}{2} \right\} + a^2 \left\{ \frac{n_0 \pi^2}{4l^2} + \frac{\alpha}{2} \right\} \cos \frac{2\pi z}{l} \right) dz. \qquad (8.7)$$

It can be seen that when the integration is carried out the third term will vanish at both limits, since it gives sin 0 and sin π. The first term will give simply $n_0 l$, the optical path length along the axis. Now if the trajectory of eqn (8.5) is really a physically possible ray of those rays which form an image of P at P′ the whole integral must be constant for varying a and equal to $n_0 l$, i.e. the second term must vanish, or,

$$\frac{n_0 \pi^2}{4 l^2} = \frac{\alpha}{2}.$$

This therefore must give the distance between successive intersections of the axis:

$$l = \pi \sqrt{\left(\frac{n_0}{2\alpha}\right)}. \tag{8.8}$$

The above analysis applies only to meridian rays and the theory for rays which are skew to the axis is considerably longer. In fact it is found that the graded index fibre will confine skew rays to a region near the axis and thus it acts in a way similar to the step index fibre.

The limiting acceptance angle for a meridian ray can be taken as that for which the ray just penetrates the region of uniform refractive index. For a ray entering at the axis this can be found by treating the continuously varying index as a sequence of thin shells of uniform index, as in Fig. 8.8. Then by applying Snell's law we find for the acceptance angle

$$\sin \theta = \sqrt{(n_0^2 - (n_0 - \alpha r_0^2)^2)}$$

which is analogous to eqn (8.2). This, however, applies only to a ray incident on the end face at the axial point. Meridian rays not incident at the axial point would have a smaller acceptance angle whereas skew rays usually have a larger acceptance angle. Thus the concept of numerical aperture for a graded index fibre is not so simple as for a step index fibre.

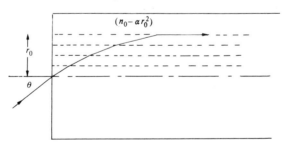

FIG. 8.8.   The limiting acceptance angle θ for a graded index fibre; the ray is calculated to be parallel to the axis when it reaches the region of uniform refractive index $n_0 - \alpha r_0^2$.

8.3. Light guides for image transport

There are three main groups of applications for light guides, (a) for transporting light simply for illumination; (b) for transporting images; and

(c) as communication channels. The first application needs no discussion as the principle is obvious; in this section we explain the use of guides for transporting images. If a bundle of fibres is made as in Fig. 8.9 and a real image is projected onto one end then it is obvious that, provided the arrangement of the fibres is kept the same at both ends, the image will appear at the other end of the bundle. Such fibre bundles are made either flexible or rigid; in the second case the individual fibres, usually of step index form, are fused together as a solid glass rod.

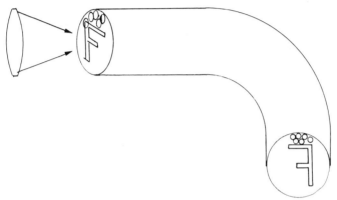

FIG. 8.9.   Image transport by a fibre bundle; the diameter of the individual fibres is not shown to scale.

In order to get the maximum image brightness the convergence angle α from the lens which projects the initial real image should equal the acceptance angle of the fibres, as given by eqn (8.2). For commercially made fibres this acceptance angle ranges between about $10°$ and $30°$; the latter value corresponds to a lens with aperture ratio $F/1$. This leads to applications where large light collecting power is essential.

Consider, for example, an oscilloscope used for recording a single very rapid event detected as, say, a voltage pulse lasting for 10^{-8} s. The trace on the oscilloscope screen must be photographed and a camera lens of very great light collecting power is needed. A typical arrangement would be as in Fig. 8.10, which shows the phosphor inside the necessarily thick faceplate of the cathode ray tube, the camera lens, and the film. It is difficult and expensive to obtain a camera lens with a large enough collecting angle to give an adequate exposure for rapidly occurring single traces and this has led to the development of the fibre-optics faceplate. Figure 8.11 shows a plate composed of short step index fibres vacuum sealed in a matrix of black glass; this plate is used as the faceplate of the cathode ray tube and the phosphor is deposited on the inner surface. Light from the electron beam trace on the

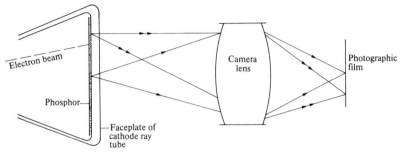

FIG. 8.10.   Taking a photograph of a single trace on an oscilloscope screen.

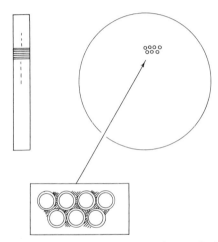

FIG. 8.11.   Structure of a fibre-optic face-plate for a cathode ray tube.

phosphor is collected with the full numerical aperture of the fibres and a piece of photographic film pressed against the outside will receive the image. With this system the light flux collected per unit area of the phosphor is proportional to the square of the numerical aperture of the fibres multiplied by a factor A which allows for the packing of the fibres. There is no distortion or other image-forming aberration and the resolving power depends only on the spacing of the fibres. For example, let the individual fibre axes be arranged in a honeycomb pattern at centre spacing d and let the diameter of the high index core of each fibre be $2a$. Then it can be seen that the factor to allow for packing is

$$A = \frac{2\pi}{\sqrt{(3)}} (a/d)^2. \tag{8.10}$$

The effect on resolving power is indicated in Fig. 8.12; a bright narrow line imaged on one side of the face plate is broadened to a width which may be as large as $d + 2a$ or as small as $2a$, depending on the position and orientation of the line relative to the fibres.

The image transport principle used in the fibre optics face plate is applied in other devices, for example as a way of eliminating field curvature in conventional lens systems and for transporting images through narrow tubes over long distances for, e.g. inspecting inaccessible cavities in machines and chemical apparatus.

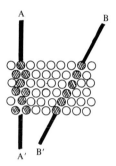

FIG. 8.12.   Resolving power of a fibre-optic face-plate; the line AA′ is broadened more than BB′ since it overlaps more fibres. In practice the arrangement of the fibres would not be as regular as in the diagram and the broadening would be irregular.

8.4. Light guides for communication

It was first suggested in 1966 that step-index light guides should be used for transmitting signals, i.e. as communication channels. The principle is, with hindsight, obvious enough, to send light down the fibre with the intensity modulated in time to give either a digital or an analogue signal. However, it was only at about that time that light sources and detectors began to be available with characteristics suitable to make such a system competitive with electrical communication along wires. In order to understand the reasons for this we have to introduce some ideas about communications.

Sounds are heard at frequencies between approximately 50 Hz and 15 kHz and they are transmitted as analogue signals, i.e. variations of voltage proportional to the instantaneous sound pressure, along telephone wires; at suitable intervals there are amplifiers, called *repeaters*, to keep the signal strength up. There would be no difficulty in modulating the output of a light-emitting diode (LED) of the kind used in hand calculators for example, at frequencies in this range, sending the light down a fibre light guide, detecting it with a photoelectric detector and then converting the electrical signal into sound in the usual way. However, this would be much more expensive and complicated than ordinary telephony. To see how light guide transmission could be competitive we note that sound covers a range of frequencies of, say, 15 kHz and this can be transmitted not directly in this frequency range but as a *modulation* of a higher frequency, the *carrier* frequency. Thus in ordinary amplitude modulation (AM) radio the carrier may have a frequency of, say, 1 MHz and the amplitude of this carrier would be varied according to the audio signal. For a carrier frequency v_0 and a single audio tone of frequency v the modulated carrier takes the form

$$V(t) = V_0(1 + m \cos 2\pi vt) \cos 2\pi v_0 t. \tag{8.11}$$

In this equation m represents the strength of the modulation. Equation (8.11) can be rewritten in the form

$$V(t) = V_0 \cos 2\pi v_0 t + \tfrac{1}{2}mV_0[(\cos 2\pi(v + v_0)t + \cos 2\pi(v - v_0)t)]$$

or in complex notation, using the convention explained in Section 1.3 that only the real part is to be taken,

$$V(t) = V_0 e^{-2\pi i v_0 t} + \tfrac{1}{2}mV_0(e^{-2\pi i(v + v_0)t} + e^{-2\pi i(v - v_0)t}) \tag{8.12}$$

We can now use our Fourier transform ideas again and say that this signal is represented in the transform domain, i.e. the temporal frequency domain, by

$$\tilde{V}(v) = V_0\delta(v_0) + \tfrac{1}{2}mV_0(\delta(v + v_0) + \delta(v - v_0)) \tag{8.13}$$

and this is shown in Fig. 8.13. The audio signals can occupy a relatively small range of frequency covering perhaps ± 15 kHz on either side of the carrier

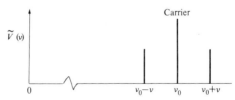

FIG. 8.13.   A carrier wave of frequency v_0, amplitude modulated at frequency v, shown in the frequency domain.

frequency v_0. This range is called the *bandwidth*. The idea of a bandwidth of temporal frequencies needed to transmit certain kinds of signals is paralleled exactly by the ideas introduced in Chapter 5 of a range of spatial frequencies needed to transmit certain detail in forming the image of an object. For television a bandwidth of order 10 MHz is needed, as can easily be seen by estimating the number of individual points in a 625 line television picture and recalling that the picture is reproduced 25 times a second; the used carrier frequencies range up to about 1000 MHz, but this probably represents the practical limit for commercial radio communication. The number of different radio or television programmes which can be broadcast simultaneously in the same area can be found by plotting their carrier frequencies and bandwidths on a frequency abscissa as in Fig. 8.14; if neighbouring bands do not overlap they can be received separately without interference or *crosstalk*, provided the receiver is tuned to pick up only the band of frequencies corresponding to the

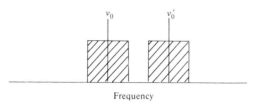

Frequency

FIG. 8.1.4. Carrier frequencies v_0 and v_0' with their accompanying required bandwidths.

desired programme. Similarly several telephone conversations can be transmitted simultaneously over one line by modulating different carrier frequencies for each; the maximum carrier frequency for a telephone line is a few MHz, so a few hundred separate telephone conversations can be carried simultaneously.

Returning to light guides, it will be recalled from Chapter 1 that visible light has frequencies in the range 4.3–7.5×10^{14} Hz. Thus by selecting suitably spaced carrier frequencies in this range a single light guide might transmit about 10^{11} separate telephone conversations or 10^8 separate television pictures simultaneously! The reality is, so far, not quite so spectacular: from what was said above it can be seen that the range of carrier frequencies and the required bandwidth are two key ideas in discussing any communication system but three further questions which arise in considering light guides as communication channels are (a) can we have enough close-spaced carriers over the frequency range of the light; (b) can a light guide transmit the signals without degrading them enough to make them unrecognizable; and (c) could the signals, if undegraded, be separated again without crosstalk? The answer to all these questions is 'No' at present. Nevertheless, light guides do offer the possibility of substantial advantages in increasing the number of separate signals carried simultaneously, as well as in other ways to be mentioned below.

Consider first the spacing of the carrier frequencies; for 10^8 separate carriers we should require the frequency of each carrier to be stable to within about ± 1 MHz; this is attainable by the most elaborate stabilization methods for a few gas lasers but by no means for 10^8 separate and arbitrarily chosen wavelengths in the visible spectrum. Equally, even supposing all these separate highly stabilized carriers could be obtained they would have to be sorted out at the receiving end of the light guide (or at intermediate switching or amplifying stations) by filters or spectroscopes of resolving power which is at present quite unattainable under the required working conditions. Thus in practice there are two possible systems, one being to modulate the light wave at a reasonable frequency, say 10^9 Hz, and use this frequency as the carrier; several carriers with suitably spaced frequencies could be used to carry simultaneously different communication channels. The second possibility is to use digital transmission and this is the method which is being most actively explored at present; in digital transmission the analogue signal, sound, television or whatever, is pre-coded into digital form by an analogue-to-digital converter. This will, for example, record the light intensity at a given point in a picture as being any of, say 64 different values in the range 0, 1, 2, ... 63 and encode it in the binary scale as a 6-bit number. A similar process is carried out with sound intensities, sampled at sufficiently close intervals of time; the signal is then transmitted as a series of 0s and 1s, as in a computer, represented as dark and light periods of the LED at the transmitting end of the light guide. Rates of modulation and detection approaching 10^9 bits per second are possible with available LEDs and photodetectors, which corresponds to an available bandwidth of about 1000 M Hz; this is still very large for a single light guide even if it is disappointing compared to the actual frequency of the light. We thus have to ask whether a fibre could transmit at such frequencies. In other words, if a single bit representing a 1 enters the fibre as in Fig. 8.15(a) will it emerge as in (b), in which case no great confusion will occur, or as in (c), when it would be

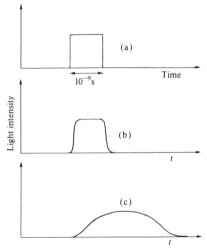

FIG. 8.15.   Degradation of a light pulse representing a single bit on transmission through a light guide: (a) input pulse, (b) slightly degraded, (c) degraded unusably.

impossible to use the light guide as a communication channel at the postulated *bit rate*.

We can formalize this in terms of the concepts of Section 5.5, but with one time dimension instead of two space dimensions. We assume the input signal to be a delta function of time, i.e. an indefinitely short burst of light, and the output from the light guide will be spread in time into the *impulse function*, the analogue of the point spread function for an optical system. The time spread of the impulse function gives a measure of the duration of signal which can be transmitted without degradation, or alternatively the Fourier transform of the impulse function gives the *transfer function*, which shows how periodic signals of different frequencies are degraded.

There are two main contributing factors to the size of the impulse function for a light guide. The first, known technically as modal dispersion, can be understood in terms of geometrical optics: in Section 8.2 we found a path for a paraxial meridian ray in a graded index fibre and it was mentioned that there could be aberrations for other rays, i.e. the optical path lengths along the fibre following different rays are not the same; but the optical path length along a ray is a measure of the time taken for light to travel the ray path (Section 2.3), so that parts of a signal travelling different ray paths will arrive at different times, producing a spread out impulse function. The second factor is simply material dispersion, the fact that the refractive index of the fibre material varies with wavelength, i.e. the speed of a light pulse varies with wavelength. Since the light used in practice has an appreciable spread of wavelengths this again contributes to the width of the impulse function.

There is a third feature of light guides which is crucial to their performance, namely *attenuation*. In a light guide only several metres long there may be a serious loss of light by absorption and scattering in the material of the guide and it is this which decides the maximum length of guide which can be used as a communication link without a repeater to restore the signal. Attenuation is expressed in decibels† per unit length. The best optical fibres obtainable at the time of writing are claimed to have attenuation as low as one or two db per kilometre. The attenuation varies with wavelength and for most fibres there is a minimum of attenuation at a wavelength of about 1.3 µm, i.e. in the near infrared region. Figure 8.16 shows, with some detail omitted, a typical attenuation curve.

Figure 8.16 suggests that to minimize the effects of attenuation we should use an infrared source of wavelength 1.3 µm. At present both LEDs and solid state lasers of different kinds which emit in the near infrared are used but there is as yet no clear best choice. Solid-state lasers of suitable design emit over a relatively narrow wavelength range of about 2 nm, so that the effect of material dispersion in a fibre is likely to be small, whereas LEDs emit over

---

† A decibel is a logarithmic unit such that 10 db is an attenuation by a factor 0.1, 20 db = 0.01, and generally n db corresponds to attenuation $10^{-0.1n}$.

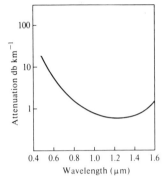

FIG. 8.16.   Attenuation curve for a fibre light guide; the curve is drawn to show the general trend for many fibre materials but it does not correspond to a particular material.

perhaps 50 to 100 nm. On the other hand LEDs are cheaper, simpler to run, and more reliable than lasers.

The effect of attenuation is to limit the length of fibre which can be used. Figure 8.17a suggests a typical digital light signal of the form 101001 to be transmitted; ignoring the effect of the impulse function of the light guide, we should naively expect the output voltage from the photoelectric detector at the exit end to have a similar form. However, in a quantum detection process (Section 1.4) the output will actually consist of individual pulses corresponding to the detection of individual photons and if the signal has been attenuated in the fibre the pulses might appear individually as in Fig. 8.17b, i.e. we should have a *noisy* output. Figure 8.17b does not show only the signal photon noise, as it is called; we have indicated also some noise at the times corresponding to zeroes in the signal; this noise is from dark current in the photodetector, i.e. stray electrons produced when there is no light, and from a variety of other sources of noise. Figure 8.17 illustrates the concept of the signal-to-noise ratio, i.e. the ratio of the mean output signal to the mean (or

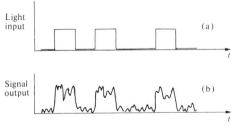

FIG. 8.17.   The effect of noise: (a) the input signal 101001, (b) the output signal from the photodetector, showing noise from various sources.

strictly root-mean-square) noise. Clearly with a signal-to-noise ratio which is too unfavourable 1s will sometimes be taken for 0s and vice versa. We can make a very simplified calculation as follows.

Let W be the light power in watts corresponding to the signal 1, let this signal last for time t and let the mean frequency of the light be v. The energy in a single 1 is Wt and this therefore contains Wt/hv photons on average. Thus $\eta Wt/hv$ photoelectrons will be produced on average for each 1, if η is the efficiency of the detector. Now it is known that the mean-square fluctuation in the number of photoelectrons produced in a given time is equal to the mean so that the root mean square variation in the number of photoelectrons produced for a 1 will be $(\eta Wt/hv)^{1/2}$ and the signal-to-noise ratio is therefore

$$S/N = \left(\frac{\eta Wt}{hv}\right)^{1/2}. \tag{8.14}$$

In this calculation we have ignored several other sources of noise. However, eqn (8.14) shows the essential features that the signal-to-noise ratio is proportional to the square root of the signal strength W and inversely proportional to the square root of the bandwidth $(1/t)$. We should note also that as given by eqn (8.14) the signal-to-noise ratio refers to current or voltage output from the photodetector. For some purposes the ratio of signal-power to noise-power obtained from the photodetector is used and then the square of the expression in eqn (8.14) would apply. Either way it can be seen that the effect of attenuation in reducing the light power W is to reduce the signal-to-noise ratio at the output from the fibre and a point is reached at which the digital signal must be refreshed by detecting the light signal, re-shaping it electronically and re-transmitting it for a further stage. The distance between such repeater stations depends on the fibre attenuation and, of course, on the strength of the input signal. This introduces the concept of coupling light guides to light sources and detectors, briefly touched on in Section 8.1.

We saw in Section 2.6 that the light transmitting power of an optical system is proportional to the square of the Lagrange invariant, a quantity sometimes called the étendue. For a step-index guide the Lagrange invariant is the product of its numerical aperture (Section 8.1) and the radius of the core. Then in order to couple a light source as efficiently as possible to the guide we must ensure that all of this étendue is used, i.e. that rays from the source enter at all angles up to the full acceptance angle and at all parts of the core cross-section. If the source is a light-emitting diode with a luminous area much greater than the core diameter this can be achieved simply by butting the source against the end of the guide. If there are objections to this because, for example, the source may be encapsulated in a glass or plastic cover we have to use a *condenser*. This is a lens or other image-forming system, which focuses an image of the source onto the end of the guide, as in Fig. 8.18. The lens must have a large enough diameter and small enough focal length to ensure that the core is covered by the source image formed by beams with convergence angles

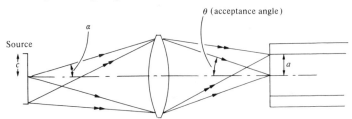

FIG. 8.18.   Imaging a light source on to the core of a step-index guide; if the image just covers the core and if the cone of rays just matches the acceptance angle of the guide the maximum possible flux is transmitted.

equal to or greater than the acceptance angle. Let the radiance (Section 2.6) of the source be B. Then as in Chapter 2 the condenser collects $\pi^2 B c^2 \alpha^2$ watts from a portion of the source of diameter $2c$. The magnification of the condenser is α/θ and if c is equal to $a\theta/\alpha$, where $2a$ is the core diameter, all this power will enter the guide. No more power can be collected by the guide from a source of radiance B and for the acceptance angle θ: for if we try to image an area of diameter $4c$ onto the core the convergence angle will become $\alpha/2$ instead of α in order to adjust the magnification, and no gain will have been achieved. A similar point was made in Section 2.6 in connection with the luminance of images and in Section 8.1 when the numerical aperture of a guide was defined.

Finally we stress that this calculation is very approximate. First, we have already noted that step index fibres transmit skew rays incident at larger angles than the formal acceptance angle θ as given by eqn (8.2), whereas on the other hand graded index fibres do *not* transmit some meridian rays at the angle given by eqn (8.9). Secondly the geometrical optics model is an approximation and to get a better calculation it would be necessary to calculate the transmission of light according to electromagnetic theory. For this, however, we refer to Midwinter (1979).

Problems

8.1.   A step index light guide has refractive index 1.54 and the cladding has index 1.52. Calculate the acceptance angle. What index would be required for the cladding to give an acceptance angle of 25°?

8.2.   Show that a guide of circular cross-section with radius a and acceptance angle θ can, within the approximations of Chapter 8, transmit a flux $\pi a^2 B \sin^2 \theta$ from a Lambertian source of radiance B.

8.3.   A graded index guide has the radial refractive index distribution $1.52 - 2r^2$ with r in mm up to a radius $r_0 = 0.2$. Find its acceptance angle and the distance between successive intersections of the axis for a meridian ray.

8.4.   A short section of the graded index guide in Problem 8.3 is to be used as an image-forming system. How long should it be if (a) an object at infinity is to be imaged at the exit end of the guide or (b) if it is to be an afocal system?

Appendix: the Fourier transform and some of its properties

Definitions

Let $f(x)$ be a function of the real variable x, single-valued and possibly complex in value. The $F(u)$, defined by

$$F(u) = \int_{-\infty}^{\infty} f(x) \exp(-i2\pi ux)\, dx, \qquad (A.1)$$

is the *Fourier transform* of $f(x)$. It can be shown that a reciprocal relationship then holds,

$$f(x) = \int_{-\infty}^{\infty} F(u) \exp(i2\pi xu)\, du, \qquad (A.2)$$

so that $f(x)$ is said to be the inverse Fourier transform of $F(u)$. We think of the functions $f(x)$ and $F(u)$ as existing in two different regions or domains, the x-domain and the u-domain, and the Fourier transformation links pairs of functions in these domains. For example, let $f(x)$ be defined by

$$f(x) = 1, \quad |x| < a/2,$$
$$f(x) = 0, \quad |x| > a/2. \qquad (A.3)$$

This is the *rectangle function*, written rect(x/a). Then by elementary integration we find for its transform,

$$F(u) = |a|\,\text{sinc}\,(au), \qquad (A.4)$$

where the *sinc function* is defined by

$$\text{sinc}\,\theta \equiv \frac{\sin \pi\theta}{\pi\theta}. \qquad (A.5)$$

This fundamental pair of functions is illustrated in Fig. A.1, and the relation between them is sometimes written in the form

$$|a|\,\text{sinc}\,au \rightleftharpoons \text{rect}\,(x/a). \qquad (A.6)$$

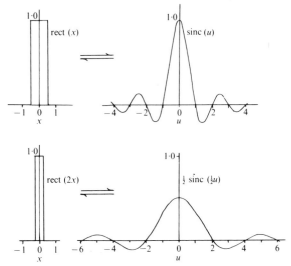

FIG. A.1.   The fundamental Fourier-transform pair, the rectangle function, and the sinc function, showing the effect of a change of scale.

This notation obscures the distinction between direct and inverse transforms but this is often unimportant in physical applications. Figure A.1 shows, through the scale factor a, how a spreading of one function (a increasing in rect (x/a)) causes the transform to be compressed along the u-axis.

Two-dimensional transforms are defined similarly in terms of functions of two variables,

$$F(u, v) = \int\int_{-\infty}^{\infty} f(x, y) \exp \{i2\pi(ux + vy)\} \, dx \, dy,$$

$$(A.7)$$

$$f(x, y) = \int\int_{-\infty}^{\infty} F(u, v) \exp \{2\pi i(xu + yv)\} \, du \, dv,$$

and the theorems given below for one-dimensional functions apply *mutatis mutandis* to two dimensions.

The definition of the Fourier transform is sometimes formulated differently: the factor 2π may be omitted from the exponent in eqns (A.1) and (A.2) and a factor $(2\pi)^{-1}$ then appears outside the integral sign in eqn (A.2), or again there may be a factor $(2\pi)^{-1/2}$ outside both integral signs. These variations seem to have roughly equal popularity with the form given in this book; there is no physical significance in the choice and it is necessary only to be consistent in usage.

The delta function

Fourier transforms can be defined for a great variety of functions, although a discussion of the conditions under which any given function can have a Fourier transform is beyond the scope of this Appendix. If $f(x)$ is a constant, say b, the integral in eqn (A.1) does not converge, and so the transform of a constant is not defined. However, if in eqn (A.6) the constant a is very large then the right-hand side is unity over this large range, and the left-hand side becomes correspondingly narrower and higher, as in Fig. A.2. In the limit as a tends to infinity the left hand side of eqn (A6) tends to an infinitely narrow and infinitely high spike. This is one way of developing a definition of the *delta*

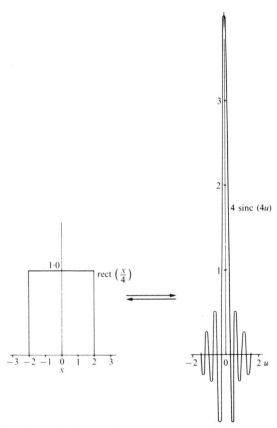

FIG. A.2.   Evolution of the delta function as the transform of a constant. As a increases the transform of rect (x/a) tends to a delta function. Other pairs of transforms can also be used to define the delta function by a similar limiting process.

function $\delta(u)$, introduced by P. A. M. Dirac. Thus we have the transform relationship

$$a \rightleftharpoons a\, \delta(u), \qquad (A.8)$$

where a is a constant. There are many situations in physics where a quantity is constant for a long time or distance, so that its transform is very sharp and narrow, almost a delta function. Thus it is the limiting process implied in eqn (A.8) which is important in physical applications of Fourier-transform theory.

Elementary properties

If we have two pairs of functions which are transforms then any linear combination can make a transform pair,

$$aF(u) + bG(u) \rightleftharpoons af(x) + bg(x). \qquad (A.9)$$

A shift of origin in one domain corresponds to multiplication by a complex exponential in the other domain,

$$f(x + a) \rightleftharpoons \exp\,(i2\pi au)F(u),$$
$$F(u - a) \rightleftharpoons \exp\,(i2\pi ax)f(x). \qquad (A.10)$$

If the above result is applied to the delta function and its transform we obtain

$$\delta(u - u_0) \rightleftharpoons \exp\,(i2\pi u_0 x), \qquad (A.11)$$

i.e. the transform of a complex exponential is a delta function with shifted origin. This result enables us to write down the transform of a periodic function. Thus if

$$f(x) = a\, \cos\, 2\pi u_0 x$$
$$= \tfrac{1}{2}a\, \exp\,(i2\pi u_0 x) + \tfrac{1}{2}a\, \exp\,(-i2\pi u_0 x)$$

we find from eqn (A.11)

$$F(u) = \tfrac{1}{2}a\, \delta(u - u_0) + \tfrac{1}{2}a\, \delta(u + u_0).$$

This can then be extended to a sum of periodic functions, e.g. a Fourier series,

$$f(x) = \Sigma a_n\, \exp\,(i2\pi nx),$$
$$F(u) = \Sigma a_n\, \delta(u - u_n).$$

A useful property of the delta function is that it encloses unit area,

$$\int_a^b \delta(u - u_0)\, \mathrm{d}u = 1, \, a < u_0 < b$$
$$= 0, u_0 < a \text{ or } b < u_0. \qquad (A.12)$$

Theorems

If $f(x)$ and $F(u)$ are a transform pair then

$$\int_{-\infty}^{\infty} |F(u)|^2 \, du = \int_{-\infty}^{\infty} |f(x)|^2 \, dx, \tag{A.13}$$

which is *Parseval's theorem.*

The *convolution* of two functions $f(x)$ and $g(x)$ is defined to be

$$f(x) \bigstar g(x) \equiv \int_{-\infty}^{\infty} f(s-x)g(x) \, dx \tag{A.14}$$

$$= \int_{-\infty}^{\infty} f(x)g(s-x) \, dx.$$

The *convolution theorem* states that the transform of the convolution of two functions is the product of their transforms,

$$f(x) \bigstar g(x) \rightleftharpoons F(u)G(u),$$

or, more explicitly,

$$\int_{-\infty}^{\infty} \exp(-\mathrm{i}2\pi su) \left\{ \int_{-\infty}^{\infty} f(s-x)g(x) \, dx \right\} ds = F(u)G(u). \tag{A.15}$$

Thus convolution in one domain corresponds to multiplication in the other.

The *autocorrelation function* of $f(x)$ is defined as

$$f(x) \stackrel{\star}{\star} f(x) \equiv \int_{-\infty}^{\infty} f(x+s)f^*(x) \, dx, \tag{A.16}$$

where $f^*(x)$ is the complex conjugate of $f(x)$; but sometimes it is more convenient to use a normalized autocorrelation function, in which case the right-hand side of eqn (A.16) is divided by the normalizing constant,

$$\int_{-\infty}^{\infty} |f(x)|^2 \, dx.$$

The *autocorrelation theorem* or *Wiener–Khintchine theorem* states that the transform of the autocorrelation of a function is the squared modulus of its transform,

$$\int_{-\infty}^{\infty} \exp(-\mathrm{i}2\pi su) \left\{ \int_{-\infty}^{\infty} f(x+s)f^*(x) \, dx \right\} ds = |F(u)|^2. \tag{A.17}$$

Optical analogies

The results given in this Appendix are mathematical expressions of many physical effects. For example, we saw in Section 3.4 that the complex amplitude in the far-field diffraction pattern of an aperture is the Fourier transform of the complex-amplitude distribution in the aperture. In a slightly different context, the same result means that the complex amplitude in the point spread function of a lens is the transform of the complex-amplitude distribution in the exit pupil of the lens (Section 5.5). Again, the complex amplitude of a plane wave striking an aperture in a plane screen normally is represented by a constant over the aperture, so that the far-field pattern tends to a delta-function shape (i.e. a narrow high peak) as the aperture gets wider. This illustrates the scaling rules given in Section 3.4.

The image of an extended object formed in incoherent illumination is the convolution of the light-intensity distribution in the point spread function of the lens with that in the object. The convolution theorem tells us that in the transform domain this corresponds to multiplying the spatial frequency distribution in the object by the optical transfer function to obtain the spatial frequency distribution in the image (Section 5.5).

The fringe function of a two-beam interferometer is the transform of the intensity distribution in the spectrum of the light, or its *power spectrum* (Section 6.2). The autocorrelation function (with time as the variable) of intensity of a polychromatic beam of light is, by the autocorrelation theorem, the transform of the square of the power spectrum of the light (Section 6.7).

Note that in the above summary we have omitted scaling and normalizing factors, etc. which would be needed for statements of the results in forms suitable for numerical calculation.

It should be noted also that Fourier transform methods can sometimes lead to dimensionally inconsistent results, particularly in the use of delta functions, since the theory itself is purely mathematical. For instance in Chapter 5 we expressed the image of an extended object as a convolution integral (eqn 5.4) and it can be seen that image and object do not have the same physical dimensions, although both are light intensities. It is possible to reformulate the theory to overcome this, at the expense of some complication, but this is not customary and it is unnecessary in our kind of optics, since we are concerned only with relative light intensities.

References and further reading

Born, M. and Wolf, E. (1975). *Principles of optics* (5th edn). Pergamon Press, Oxford.

Ingram, D. J. E. (1973). *Radiation and quantum physics* (OPS 3). Clarendon Press, Oxford.

Lipson, H. S. (ed.) (1972). *Optical transforms*. Academic Press, New York.

Midwinter, J. E. (1979). *Optical fibres for transmission*. John Wiley, New York.

Robinson, F. N. H. (1973). *Electromagnetism* (OPS 1). Clarendon Press, Oxford.

Svelto, O. (1976). *Principles of lasers* (trans. D. C. Hanna). Heyden, London.

Welford, W. T. (1962). *Geometrical optics; optical instrumentation*. North-Holland, Amsterdam.

—— (1974). *Aberrations of the symmetrical optical system*. Academic Press, London.

For descriptions of many of the classical interference and diffraction experiments see, e.g., R. W. Ditchburn (1976). *Light*. Academic Press, London or R. S. Longhurst (1973). *Geometrical and physical optics* (3rd edn). Longmans, London.

Answers to numerical problems

In some answers only one or two significant figures are given. This indicates that only that precision is physically meaningful, or that the initial data have only that precision.

1.1. 300 m, 300 mm, 300 µm, 300 nm.
1.6. 50 µm, 6×10^{12} Hz.
1.7. Curve 2, 10 per cent; curve 3, 5 per cent.
2.9. $\frac{5}{3}$, -1, $\frac{1}{3}$.
2.10. Focal length $r/2$; image positions -50 mm, 100 mm, 66.7 mm; magnifications 2, -1, $-\frac{1}{3}$.
2.11. (a) $\frac{1}{4\pi}$ W mm^{-2}sr^{-1} (in this solution it is assumed that the total area is 10 mm^2 and the filament radiates uniformly in all directions) (b) 0.03 W.
2.13. 101.75°.
2.14. 30 mm × 2.5.
3.1. 570 m.
3.2. 140 mm^{-1}.
3.6. 0.8 nm.
3.8. 10 m.
3.9. 2 mm.
3.11. 0.5 km.
4.1. (a) 0.985; 0.707; 0.035. (b) 0.970; 0.500; 0.00122.
4.2. (a) 1; (b) 4; (c) 1.987.
4.3. 56.3°, 58.0°, 62.2°.
4.5. 3.4 µm.
5.1. 6.7×10^{-6} rad, 6.7×10^{-7} rad, 1.3×10^{-7} rad (the wavelength is taken as 550 nm).
5.2. 7 µm in diameter; 2×10^{-3} rad mm^{-1}.
5.3. 7 mm.
5.5. 0.5 µm, 500.
5.6. (a) NA 0.25, × 125, (b) NA 1.3, × 1000. (The suggested answers give NA and magnification about 4 times larger than needed for resolution of the dimensions given).
6.1. (a) 0.546 mm, (b) 0.13 mm; (c) the fringe contrast would decrease.
6.2. 0.04 arcsec.
6.5. 6.5. $\delta\lambda = 0.002$ nm; $\delta v = 1.5 \times 10^9$ Hz.
6.7. 200 nm to 400 nm.
6.8. 63.5°.
6.9. 5300.
6.10. $F = 30$, $T_{min} = 0.0028$.
6.11. 6×10^5, 0.025 nm.
7.1. 12 mm diameter.

7.2. 6 μm.

7.4. (a) 7.9; (b) 5.6; (c) 0, but fringes will appear if the surface is viewed obliquely.

7.5. The beam intersect at $\pm 18.45°$ to the normal to the grating; 3160 rulings per millimetre.

8.1. $14.32°$. $n_0 = 1.481$.

8.3. $29.12°$. 1.937 mm.

8.4. (a) 0.968 mm; (b) 1.937 mm, or any multiple of this distance.

Index

Physical constants and conversion factors

Avogadro constant	L or N_A	6.022×10^{23} mol^{-1}
Bohr magneton	μ_B	9.274×10^{-24} J T^{-1}
Bohr radius	a_0	5.292×10^{-11} m
Boltzmann constant	k	1.381×10^{-23} J K^{-1}
charge of an electron	e	-1.602×10^{-19} C
Compton wavelength of electron	$\lambda_c = h/m_e c = 2.426 \times 10^{-12}$ m	
Faraday constant	F	9.649×10^4 C mol^{-1}
fine structure constant	$\alpha = \mu_0 e^2 c/2h = 7.297 \times 10^{-3} (\alpha^{-1} = 137.0)$	
gas constant	R	8.314 J K^{-1} mol^{-1}
gravitational constant	G	6.673×10^{-11} N m^2 kg^{-2}
nuclear magneton	μ_N	5.051×10^{-27} J T^{-1}
permeability of a vacuum	μ_0	$4\pi \times 10^{-7}$ H m^{-1} exactly
permittivity of a vacuum	ε_0	8.854×10^{-12} F m^{-1} $(1/4\pi\varepsilon_0 = 8.988 \times 10^9$ m F$^{-1})$
Planck constant	h	6.626×10^{-34} J s
(Planck constant)/2π	$\hbar$	1.055×10^{-34} J s $= 6.582 \times 10^{-16}$ eV s
rest mass of electron	m_e	9.110×10^{-31} kg $= 0.511$ MeV/c^2
rest mass of proton	m_p	1.673×10^{-27} kg $= 938.3$ MeV/c^2
Rydberg constant	$R_\infty = \mu_0^2 m_e e^4 c^3/8h^3 = 1.097 \times 10^7$ m^{-1}	
speed of light in a vacuum	c	2.998×10^8 m s^{-1}
Stefan–Boltzmann constant	$\sigma = 2\pi^5 k^4/15h^3 c^2 = 5.670 \times 10^{-8}$ W m^{-2} K^{-4}	
unified atomic mass unit (^{12}C)	u	1.661×10^{-27} kg $= 931.5$ MeV/c^2
wavelength of a 1 eV photon		1.243×10^{-6} m

1 Å $= 10^{-10}$ m; 1 dyne $= 10^{-5}$ N; 1 gauss (G) $= 10^{-4}$ tesla (T);
$0°$C $= 273.15$ K; 1 curie (Ci) $= 3.7 \times 10^{10}$ s^{-1};
1 J $= 10^7$ erg $= 6.241 \times 10^{18}$ eV; 1 eV $= 1.602 \times 10^{-19}$ J; 1 cal$_{th} = 4.184$ J;
$\ln 10 = 2.303$; $\ln x = 2.303 \log x$; e $= 2.718$; $\log$ e $= 0.4343$; $\pi = 3.142$